Safety, Security, and Reliability of Robotic Systems

Safety, Security, and Reliability of Robotic Systems

Algorithms, Applications, and Technologies

Edited by
Brij B. Gupta and Nadia Nedjah

CRC Press
Taylor & Francis Group
Boca Raton London New York

CRC Press is an imprint of the
Taylor & Francis Group, an **informa** business

First edition published 2021

by CRC Press
6000 Broken Sound Parkway NW, Suite 300, Boca Raton, FL 33487-2742
and by CRC Press

2 Park Square, Milton Park, Abingdon, Oxon, OX14 4RN

CRC Press is an imprint of Taylor & Francis Group, LLC

ISBN: 9780367339463 (hbk)
ISBN: 9781003031352 (ebk)

Typeset in Sabon
by codeMantra

Dedicated to my family and friends for their constant support during the course of this book.

–B. B. Gupta

Dedicated to my family and friends for their constant support during the course of this book.

–Nadia Nedjah

Contents

Preface

Robotic systems are characterized by the automation of applications to improve productivity and efficiency in terms of the number of human or non-human resources required including cost and time. These systems are involved in performing tasks including data collection, control applications, and are often mobile. With the increasing demand of robots for industrial or domestic usage, it becomes indispensable to ensure their safety, security, and reliability. This book will provide an overview of robotic systems and their applications, and how their safety, security, and reliability can be achieved using different methodologies. This book will begin with discussing the evolution of robotic systems, some industrial statistics, and forecasts. It will then focus on the preliminary concepts in designing robotic systems including the mathematical models and learning mechanisms along with different underlying technologies. Successively, it will discuss the safety-related parameters of these systems. Then, this book will cover security attacks and related countermeasures, and how to establish reliability in these systems. It will also discuss various applications of these systems. The contribution of this book will be useful to the system designers, software developers, researchers, scholars, students, and security professionals in the field.

The proposed book will guide readers to follow their own paths of learning yet structured in distinctive modules that permit flexible reading. It will be a well-informed, revised, and comprehensible educational and informational book that addresses not only professionals but also students interested in safety, security, and reliability of robotic systems. This book will be a valuable resource for readers at all levels interested in safety, security, and reliability of robotic systems, serving also as an excellent reference in cyber security for professionals in this fast-evolving and critical field. The contents of this book will be very refreshing, informative, containing an impressive collection of up-to-date safety, security, reliability, and trust issues, and analysis of robotic systems. The chapters in this book present the fundamental concepts of robotic systems as well as in-depth coverage of theory, challenges, their risk factors, attacks, and solutions to cope with them. It explores practical solutions to a wide range of safety, security,

and reliability challenges of robotic systems. Advanced security tools, algorithms, techniques, and real-life case studies make this an ideal book.

This book contains chapters dealing with different aspects of Safety, Security, and Reliability of Robotic Systems. These include Overview, Evolution, Forecasts, Components, and Working of Robotic Systems, Collaborative Robotic Environment, Artificial Intelligence (AI) in Robotic Systems, Machine Learning in Robotic Systems, Computer Vision, Imitation Learning, and Self-Supervised Learning in Robotic Systems, Soft Computing for Robotic Systems, Security Measures in Robotic Systems, Control, Safety, Security, and Reliability in Robotic Systems, Human Factors and System Handling of Robotic Systems, Application Areas of Robotic Systems, Cloud and Social Media Analysis in Robotics, Robotics in Healthcare Systems, Robotic Systems in Unmanned Vehicular Applications, Standards, Policies and Government Regulations for Robotic Systems.

Specifically, this book contains discussion on the following topics:

- Robotic Systems Overview and Its Application in Modern Times
- Development of a Humanoid Robot's Head with Facial Detection and Recognition Features Using Artificial Intelligence
- Detecting DeepFakes for Future Robotic Systems
- Nanoscale Semiconductor Devices for Reliable Robotic Systems
- Internet of Things for Smart Gardening and Securing Home from Fire Accidents
- Deep CNN-Based Early Detection and Grading of Diabetic Retinopathy Using Retinal Fundus Images
- Vehicle Detection Using Faster R-CNN
- Two-Phase Authentication and VPN-Based Secured Communication for IoT Home Networks
- An Efficient Packet Reachability-Based Trust Management Scheme in Wireless Sensor Networks
- Spatial Domain Steganographic Method Detection Using Kernel Extreme Learning Machine
- An Efficient Key Management Solution to Node Capture Attack for WSN
- Privacy Preservation and Authentication Protocol for BBU-Pool in C-RAN Architecture
- Threshold-based Technique to Detect a Black Hole in WSN
- Credit Card Fraud Detection by Implementing Machine Learning Techniques
- Authentication in RFID Scheme Based on Elliptic Curve Cryptography
- Iris-Based Privacy-Preserving Biometric Authentication Using NTRU Homomorphic Encryption

MATLAB® is a registered trademark of The MathWorks, Inc. For product information,
Please contact:

The MathWorks, Inc.
3 Apple Hill Drive
Natick, MA 01760-2098 USA
Tel: 508-647-7000
Fax: 508-647-7001
E-mail: info@mathworks.com
Web: www.mathworks.com

Acknowledgment

Many people have contributed greatly to this book on *Safety, Security, and Reliability of Robotic Systems: Algorithms, Applications, and Technologies*. We, the editors, would like to acknowledge all of them for their valuable help and generous ideas in improving the quality of this book. With our feelings of gratitude, we would like to introduce them in turn. The first mention is the authors and reviewers of each chapter of this book. Without their outstanding expertise, constructive reviews, and devoted efforts, this comprehensive book would become something without contents. The second mention is the CRC Press-Taylor & Francis Group staff, especially Ms. Gabriella Williams and her team for their constant encouragement, continuous assistance, and untiring support. Without their technical support, this book would not be completed. The third mention is the editor's family for being the source of continuous love, unconditional support, and prayers not only for this work but throughout our lives. Last but far from least, we express our heartfelt thanks to the Almighty for bestowing over us the courage to face the complexities of life and complete this work.

June 2020
Brij B. Gupta
Nadia Nedjah

Acknowledgment

Editors

Dr. Brij B. Gupta received his PhD degree from the Indian Institute of Technology Roorkee, India, in the area of Information and Cyber Security. In 2009, he was selected for the Canadian Commonwealth Scholarship Award by the Government of Canada. He has published more than 300 research papers (SCI/SCIE Indexed Papers: 125+) in international journals and conferences of high repute including IEEE, Elsevier, ACM, Springer, Wiley, Taylor & Francis, and Inderscience. He has visited several countries such as Canada, Japan, Australia, China, Spain, the UK, the USA, Thailand, Hong Kong, Italy, Malaysia, and Macau, to present his research work. His biography was selected and published in the 30th Edition of Marquis Who's Who in the World, 2012. In addition, he has been selected to receive the "2017 Albert Nelson Marquis Lifetime Achievement Award" and the "2017 Bharat Vikas Award" by Marquis Who's Who in the World, USA, and ISR, India, respectively. He also received the "Sir Visvesvaraya Young Faculty Research Fellowship Award" (INR 3.7 Millions) in 2017 from the Ministry of Electronics and Information Technology, Government of India. Recently, he has received the "2018 Best Faculty Award for Research Activities" and the "2018 Best Faculty Award for Project and Laboratory Development" from National the Institute of Technology (NIT), Kurukshetra, India, and the "2018 RULA International Award for Best Researcher." He also received the "2019 Best Faculty Award" from NIT and "Outstanding Associate Editor of 2017" for *IEEE Access* and the "2020 ICT Express Elsevier Best Reviewer Award." His Google Scholar h-index is 48 (including i10 index: 151) with around 7740 citations for his work. He is also working as a principal investigator of various R&D projects sponsored by various funding agencies of Government of India. He is serving/served as associate editor (or guest editor) of *IEEE Transactions on Industrial Informatics*, *IEEE Access*, *ACM TOIT*, *IEEE IoT*, *FGCS*, *Applied Soft Computing*, *Soft Computing*, *ETT*, Wiley, *PPNA*, *Enterprise Information System*, *IJES*, *IJCSE*, *IJHPSA*, *IJICA*, *IJCSE*, *CMC*, *Tech Science*, *Journal of Computational Science*, *Journal of Cleaner Production*, *Technology Forecasting and Social Changes*, etc.

Moreover, he is also editor-in-chief of the *International Journal of Cloud Applications and Computing* (IJCAC) and *International Journal of Software Science and Computational Intelligence* IJSSCI, IGI Global, USA. Moreover, he is serving as managing editor of International Journal of High Performance Computing and Networking (IJHPCN), Inderscience. He also leads a book series on *Security, Privacy, and Trust in Mobile Communications*, CRC Press, Taylor & Francis, and a book series on *Cyber Ecosystem and Security*, CRC Press, Taylor & Francis, as editor-in-chief. He also served as TPC chair of 2018 IEEE INFOCOM: CCSNA in the USA. Moreover, he served as publicity chair of the 10th NSS 2016, 17th IFSA-SCIS 2017, and 29th ICCCN 2020 which were held in Taiwan, Japan, and the USA, respectively. He is also founder chair of FISP and ISCW workshops organized in different countries every year. He is serving as organizing chair of the Special Session on Recent Advancements in Cyber Security (SS-CBS) in the IEEE Global Conference on Consumer Electronics (GCCE), Japan, every year since 2014. He has also served as a technical program committee (TPC) member or in other capacities of more than 100 international conferences worldwide. He received the Outstanding Paper Awards in both regular and student categories in the 5th IEEE Global Conference on Consumer Electronics (GCCE) in Kyoto, Japan, October 7–10, 2016. In addition, he received the "Best Poster Presentation Award" and the "People Choice Award" for poster presentation in CSPC-2014, August 2014, Malaysia. He served as jury in All IEEE-R10 Young Engineers' Humanitarian Challenge (AIYEHUM-2014), 2014. He also received the "WIE Best Paper Award," in the 8th IEEE GCCE-2019, Osaka, Japan, October 15–18, 2019. He is a senior member of IEEE, Member ACM, SIGCOMM, SDIWC, Internet Society, Institute of Nanotechnology, Life Member, International Association of Engineers (IAENG), Life Member, International Association of Computer Science and Information Technology (IACSIT). He was also a visiting researcher with Yamaguchi University, Yoshida, Japan, Deakin University, Geelong, Australia, and the Swinburne University of Technology, Melbourne, Australia, January 2015 and 2018, July 2017, and March to April 2018, respectively. Moreover, he was also a visiting professor in the University of Murcia, Spain, University of Naples Federico II, Italy, and University of Salerno, Italy, in June to July 2018. He was also a visiting professor with Temple University, Philadelphia, PA, USA, and Staffordshire University, UK, June 2019 and July 2020, respectively. At present, he is group lead, Information and Cyber Security, and assistant professor in the Department of Computer Engineering, National Institute of Technology, Kurukshetra, India. His research interest includes information security, cyber security, cloud computing, web security, mobile/smartphone, intrusion detection, IoT security, AI, social media, computer networks, and phishing.

Dr. Nadia Nedjah graduated in 1987 in Systems Engineering and Computation and in 1990 obtained an M.Sc. degree also in Systems Engineering and Computation from the University of Annaba, Algeria. Since 1997, she holds a PhD degree from the Institute of Science and Technology, University of Manchester, UK. She joined as associate professor in the Department of Electronics Engineering and Telecommunications of the Engineering Faculty of the State University of Rio de Janeiro, Brazil. Between 2009 and 2013, she was the head of the Intelligent System research area in the Electronics Engineering Post-Graduate Programme of the State University of Rio de Janeiro. She is the founder and editor-in-chief of the_*International Journal of High Performance System Architecture* and the *International Journal of Innovative Computing Applications*, both published by Inderscience, UK. She published three authored books: *Functional and Re-writing Languages*, *Hardware/Software Co-design for Systems Acceleration* and *Hardware for Soft Computing vs. Soft Computing for Hardware*. She (co-)guest-edited more than 20 special issues for high-impact journals and more than 45 organized books on computational intelligence-related topics, such as *Evolvable Machines*, *Genetic Systems Programming*, *Evolutionary Machine Design: Methodologies and Applications*, and *Real-World Multi-Objective System Engineering*. She (co-)authored more than 120 journal papers and more than 200 conference papers. She is an associate editor of more than ten international journals, such as the Taylor & Francis's *International Journal of Electronics*, Elsevier's *Integration, the VLSI Journal*, and *Microprocessors and Microsystems*, and IET's *Computer & Digital Techniques*. She organized several major conferences related to computational intelligence, such as the seventh edition of *Intelligent Systems Design and Application* and the fifth edition of *Hybrid Intelligent Systems*. She was also one of the founders of the International Conference on Adaptive and Intelligent Systems. (More details can be found at her homepage: http://www.eng.uerj.br/~nadia/english.html.)

Contributors

Sohaib Abdullah did Bachelor's and Master's degree from Manarat International University and Dhaka University, Bangladesh, respectively. Currently, he is working with Department of Computer Science and Engineering, Manarat International University, Bangladesh.

Mr. Abhay Kumar Agrahari is currently pursuing Ph.D. at Indian Institute of Information Technology-Allahabad in the Department of Information Technology (IT). His areas of research include RFID, RFID security, cryptography, and IOT.

Dr. Priyanka Ahlawat is an Assistant Professor in Computer Engineering Department, National Institute of Technology, Kurukshetra, Haryana, India. She obtained her Ph.D. degree from the National Institute of Technology. Her interest areas are key management, wireless sensor networks security, and cryptography. She has several years of experience working in academia, teaching, and research. She has published more than 35 research papers in various conferences, workshops, and international journals of repute including IEEE. She has organized several international conferences and short-term courses in India. She is a member of various professional societies like IEE, ACM, and CSI. She has published a book from IGI Global USA.

Rangel Arthur. He holds a degree in Electrical Engineering from the Paulista State University Júlio de Mesquita Filho (1999), a Master's degree in Electrical Engineering (2002), and a Ph.D. in Electrical Engineering (2007) from the State University of Campinas. Over the years from 2011 to 2014 he was Coordinator and Associate Coordinator of Technology Courses in Telecommunication Systems and Telecommunication Engineering of FT, which was created in its management. From 2015 to 2016 he was Associate Director of the Technology (FT) of Unicamp. He is currently a lecturer and advisor to the Innovation Agency (Inova) of Unicamp. He has experience in the area of Electrical Engineering, with emphasis on Telecommunications Systems, working mainly on the following topics: computer vision, embedded systems, and control systems.

Shaveta Chutani was born in Ludhiana, India, in 1977. She received her M.C.A. degree from IGNOU, New Delhi, India in 2006 and M.Tech. degree in Computer Applications from I.K. Gujral Punjab Technical University, Jalandhar, India in 2012. She is currently pursuing her Ph.D degree in Computer Applications from I.K. Gujral Punjab Technical University, Jalandhar, India.

She has over 12 years of experience as Assistant Professor with the Computer Applications Department in various educational institutions. Her research interests include information security, steganography and steganalysis.

Dr. Veerendra Dakulagi received his B.E. and M.Tech. degrees from Visvesvaraya Technological University, Belagavi, India, in 2007 and 2011, respectively. He received his Ph.D. degree in Adaptive Antennas from the same university in 2018. From 2010 to 2017, he was an Asst. Professor in Dept. of E&CE, Guru Nanak Dev Engineering College, Bidar, India. Since March 2017, he has been an Associate Professor and R&D Dean of the same institute. His research interests are in the area of signal processing and communications, and include statistical and array signal processing, adaptive beamforming, spatial diversity in wireless communications, multiuser, and MIMO communications. He has published over 40 technical papers (including IEEE, Elsevier, Springer, and Taylor and Francis) in these areas. Dr. Veerendra Dakulagi is a member of Institute of Electronics and Telecommunication Engineers (IETE) and currently serves as an editorial board member of *Journal of Computational Methods in Sciences and Engineering*, *Journal of Communication Engineering and Systems*, *Journal of Electronics Engineering*, and *International Journal of Scientific Research and Development*. Currently he is also working as a Post-Doctoral Fellow (PDF) at the Faculty of Electronics & Communication Engineering, Lincoln University College, Petaling Jaya, Malaysia.

Prof. Mayank Dave obtained his M.Tech. degree in Computer Science and Technology from IIT Roorkee, India in 1991 and Ph.D. from the same institute in 2002. He is presently working as an Associate Professor in the Department of Computer Engineering at NIT Kurukshetra, India with more than 19 years of experience in academic and administrative affairs at the institute. He has published approximately 85 research papers in various international/national journals and conferences. He has coordinated several projects and training programs for students and faculty. He has delivered a number of expert lectures and keynote addresses on different topics. He has guided four PhDs and several M.Tech. dissertations. His research interests include peer-to-peer computing, pervasive computing, wireless sensor networks, and database systems.

P. P. Deepthi received M.Tech. degree in Instrumentation from Indian Institute of Science, Bangalore in 1997 and Ph.D. from National Institute

of Technology, Calicut in 2009 in the field of Secure Communication. She is currently Professor in the Department of Electronics and Communication Engineering, National Institute of Technology, Calicut. Her current interests include cryptography, information theory and coding theory, multimedia security, and secure signal processing.

E. Devi received her B.Tech. degree in Electronics and Communication Engineering from Madanapalle Institute of Technology and Science, Madanapalle. She is currently pursuing M.Tech. in Signal Processing at National Institute of Technology, Calicut. Her current interests include signal processing, image processing, cryptography, and machine learning.

Dr. Chander Diwaker is presently working as an Assistant Professor in the Department of Computer Engineering, University Institute of Engineering & Technology, Kurukshetra University, Kurukshetra. He has nearly 13years of teaching experience. He obtained his M.Tech. degree from YMCA Institute of Engineering (now J.C. Bose University of Science and Technology), Faridabad in 2004. He obtained his Ph.D. degree from the School of ICT, Gautam Buddha University, Greater Noida in 2018. His current research interests mainly include wireless networks, software engineering, internet of things (IoT), and computer engineering. He has published 50 research papers in referred international journals and international conferences.

Reinaldo Padilha França. Master in Electrical Engineering from the State University of Campinas - UNICAMP (2018), Computer Engineer (2014), and Logistics Technology Management (2008) from the Regional University Center Espírito Santo of Pinhal. Has knowledge in programming and development in C/C++, Java, and .NET languages, from projects and activities. Natural, dynamic, proactive, communicative, and creative self-taught. He is currently developing research projects concerning pre-coding and bit coding in AWGN channels and digital image processing with emphasis on blood cells. Such research derived a computer program record (2019). He is a researcher at the Laboratory of Visual Communications (LCV). He serves at the Brazilian Technology Symposium (BTSym) as a member of the Organizational and Executive Committee and as a member of the Technical Reviewers Committee. Has interest and affinity in the areas of technological and scientific research, teaching, digital image processing, and Matlab.

Amit Kumar Gautam received his B.Tech in Information Technology in 2008 from BIET Jhansi (UP), India and M.Tech degree in Information Technology specialization in Software Engineering from IIIT Allahabad, India. Since July 2017, he has been a Research Scholar under UGC NET/JRF in the Department of Computer Science and Engineering, MMM University of Technology, Gorakhpur. His area of interests includes computer networks, Wireless sensor network and WSN Security.

Anjali Goyal received her Bachelor degree in Electronics in 1993 from Kurukshetra University and Master degree in Computer Applications in 1996 from Panjab University, Chandigarh. She received her Ph.D degree from Punjab Technical University, Jalandhar, India in 2013. Her academic achievement includes University Merit position in Graduation. Presently she is working as Assistant Professor in Department of Computer Application at Guru Nanak Institute of Management and Technology, Ludhiana affiliated to PTU. Earlier she has served in various colleges of Kurukshetra University and Punjab Technical University. She has a teaching experience of 18 years. Her research interests include Content Base Image Retrieval, Digital Watermarking and Pattern recognition. She has a number of International journal and conference publications to her credit. She is reviewer of many reputed International journals. She is also guiding many Ph.D students.

Dr. Koppala Guravaiah obtained his graduate degree in Computer Science and Engineering in 2009 from Audisankara College of Engineering and Technology, J.N.T.U. Ananthapur and postgraduate degree in Computer Science in 2011 from Sree Vidyanikethan Engineering College, J.N.T.U. Ananthapur, Andhra Pradesh, India. He completed his Ph.D. from National Institute of Technology (NIT), Tiruchirappalli, Tamil Nadu, India. Currently, he is working as a Senior Assistant Professor, Department of Computer Science & Engineering, Madanapalle Institute of Technology & Science, Madanapalle, Andhra Pradesh, India. His research interests include data collection routing in Wireless Sensor Networks, Internet of Things, and Mobile Ad Hoc Networks.

Md Mahedi Hasan is working with Institute of Information and Communication Technology, Bangladesh University of Engineering and Technology, Bangladesh.

Yuzo Iano. BS (1972), Master's degree (1974) and a Ph.D. degree (1986) in Electrical Engineering from the State University of Campinas, Brazil. Since then he has been working in the technological production field, with 1 patent granted, 8 patent applications filed, and 36 projects completed with research and development agencies. Successfully supervised 29 doctoral theses, 49 master's dissertations, 74 undergraduate, and 48 scientific initiation work. He has participated in 100 master's examination boards, 50 doctoral degrees, author of 2 books, and more than 250 published articles. He is currently a Professor at the State University of Campinas, Brazil, Editor-in-Chief of the SET International Journal of Broadcast Engineering, and General Chair of the Brazilian Symposium on Technology (BTSym). He has experience in Electrical Engineering, with knowledge in Telecommunications, Electronics and Information Technology, mainly in the field of audio-visual communications and data.

Mr. Anik Islam was born in 1992. He received the B.Sc. in software engineering and M.Sc. degrees in computer science from American International University-Bangladesh (AIUB), Dhaka, Bangladesh, in 2014 and 2017, respectively. He is currently working toward the PhD degree with the WENS Laboratory, Kumoh National Institute of Technology, Gumi, South Korea. He has more than 5 years of experience of working in the software development field. He has participated in various software competitions with good achievements. His major research interests include blockchain, internet of things, unmanned aerial vehicle, social internet of things, edge computing, and distributed system.

Mr. Md Mofijul Islam is currently pursuing Ph.D. Degree in System and Information Engineering at University of Virginia. He received the B.S. and M.S. degrees in computer science and engineering from the Department of Computer Science and Engineering, University of Dhaka, Bangladesh. He was a Lecturer at the Department of Computer Science and Engineering, University of Dhaka (Sep 2017–Aug 2019). He also was a Lecturer at the Department of Computer Science and Engineering, United International University (Jan 2015–Sep 2017). He was a Software Engineer with Tiger It Ltd (Apr 2014–Jul 2014). He is also involved in programming training, mobile apps, and different software contest team-building activities. His research interests include Artificial Intelligence, Multimodal Learning, Self-Supervised Learning, Human-Robot Interaction, and Optimization.

Sheikh Muhammad Saiful Islam did M.Pharm in Pharmaceutical chemistry and B.Pharm (Honors) from Dhaka University, Bangadesh. Currently he is working with Department of Pharmacy, Manarat International University, Bangladesh.

Mr. Jiun Hao Koh enrolled in the Bachelor of Engineering (Honours) Electronic Engineering in Universiti Tunku Abdul Rahman in 2016. He is currently in his final academic year and is expected to graduate with a first-class degree.

Rakesh Kumar is a Professor in the Department of Computer Science and Engineering at MMM University of Technology, Gorakhpur (UP), India. He received his PhD from Indian Institute of Technology, Roorkee, India in 2011. He has supervised a large number of M Tech dissertations, guided four PhD students and also guiding several M.Tech and PhD students. He has published a large number of research papers in the international journals, conferences and book chapters of high repute. His main interests are wireless sensor networks, mobile distributed computing, MANET-Internet integration, network security and Cloud Computing etc. He has successfully completed major research projects funded by University Grants Commission (UGC) and AICTE. He is a Member of IEEE, The Institution of Engineers (Fellow), Institution of Electronics

and Telecommunication Engineers (F), Life Member of Indian Society for Technical Education, Computer Society of India (India), International Association of Engineers (IAENG) and Member, ACM.

Mr. Shailesh Kumar is a Ph.D. scholar in the Department of Computer Engineering, NIT Rourkela Odisha, India. He has completed his M.Tech. degree from NIT Rourkela in 2018. His areas of interests includes wireless sensor networks (WSNs) and internet of things (IoT). He is presently working on C-IoT security and cyber security.

Ir. Dr. Siu Hong Loh is an Assistant Professor in the Faculty of Engineering and Green Technology at Universiti Tunku Abdul Rahman (UTAR) where he has been a faculty member since 2011. Loh completed his Ph.D. at University College Dublin, Ireland and his Master's study at Queen's University of Belfast, UK. He completed his undergraduate studies at Sheffield Hallam University. His research interests lie in the area of atomic force microscopy, ranging from theory to design and implementation. He has collaborated actively with researchers in several other disciplines of electronic engineering, particularly in artificial intelligence (AI), very large-scale integrated circuit (VLSI) design, and semiconductor manufacturing process. Loh has served as the faculty representative in the University Examination Disciplinary Committee (EDC) since 2013. He has also been the department representative in the undergraduate Final Year Project Committee and the Department Curriculum Development Committee (DCDC) since 2015. Loh is a corporate member of the Institution of Engineers, Malaysia (IEM), and he is also a Professional Engineer who is registered with the Board of Engineers Malaysia (BEM).

Mr. Byomakesh Mahapatra is currently pursuing his Ph.D. degree in Computer Science and Engineering at National Institute of Technology (NIT), Rourkela, Odisha, India. He earned his Master's degree (M.Tech.) in Electronics Engineering from Tezpur University, Assam, India in 2012. His research interests include design of application-specific radio access network platform for next-generation wireless and cellular network. Mr. Mahapatra has published almost 18 research articles in the field of cellular communication and networking. Presently he is involved in the design and development of multi-RAT Cloud Radio Access Network platform. He is a student member of IEEE and IEI.

Mr. Ashik Mahmud is currently is pursuing his Masters degree at Hochschule Rhein-Waal University, Germany. Earlier he has worked as a web developer at Composis Blades Inc. in Bangladesh as software engineer from 2018 to 2019. His major research interest includes web development, web engineering, web Security etc. He has several publication on international conferences.

Mr. Md Masuduzzaman is currently pursuing his Ph.D in IT convergence engineering at Kumoh National Institute of Technology, Gumi, South Korea. Previously he has worked as a Lecturer at American International University-Bangladesh for 4 years (2015–2019). He is in study leave now to complete his PhD program. His major research interests include Blockchain, Internet of Things (IoT), Unmanned Aerial Vehicle (UAV), Edge Computing, Machine Learning, Cryptography and Network Security. He has several publications on International Journal and Conferences across the world.

Dr. Melody Moh obtained her M.S. and Ph.D., both in Computer Science, from University of California, Davis. She joined San Jose State University in 1993, and has been a Professor since August 2003. Her research interests include cloud computing, mobile, wireless networking, security/privacy for cloud and network systems, and machine learning/deep learning applications. She has received over 550K dollars of research grants from both NSF and industry, has published over 150 refereed papers in international journals, conferences, and as book chapters, and has consulted for various companies.

Dr. Teng-Sheng Moh received his Ph.D. in Computer Science from University of California, Davis. He is currently a faculty member in the Department of Computer Science, San Jose State University, which he joined in August 2005. Before that, he spent 15 years in various software companies in Silicon Valley. His papers, coauthored with students, have won the best paper/poster paper awards at 2014 IEEE International Conference on High Performance Computing and Simulation (HPCS), 2019 ACM Annual Southwest Conference (ACMSE), 2020 International Conference on Ubiquitous Information Management and Communication (IMCOM), and most recently the first-place award in the Video Virtual Try-on Challenge, at 2020 Towards Human-Centric Image/Video Synthesis CVPR (Computer Vision and Pattern Recognition) Workshop.

Ana Carolina Borges Monteiro. She is a Ph.D. student at the Faculty of Electrical and Computer Engineering (FEEC) at the State University of Campinas - UNICAMP, where she develops research projects regarding health software with emphasis on the development of algorithms for the detection and counting of blood cells through processing techniques. digital images. These projects led in 2019 to a computer program registration issued by the INPI (National Institute of Industrial Property). She holds a Master's degree in Electrical Engineering from the State University of Campinas - UNICAMP (2019) and graduated in Biomedicine from the University Center Amparense - UNIFIA with a degree in Clinical Pathology - Clinical Analysis (2015). In 2019, he acquired a degree in Health Informatics. Has to experience in the areas of Molecular Biology and management with research animals. Since 2017, she has been a

researcher at the FEEC/UNICAMP Visual Communications Laboratory (LCV) and has worked at the Brazilian Technology Symposium (BTSym) as a member of the Organizational and Executive Committee and as a member of the Technical Reviewers Committee. In addition, she works as a reviewer at the Health magazines of the Federal University of Santa Maria (UFSM - Brazil), Medical Technology Journal MTJ (Algeria), and Production Planning & Control (Taylor & Francis). Interested in: digital image processing, hematology, clinical analysis, cell biology, medical informatics, Matlab, and teaching.

Pallavi S. Mote is a Master's student at the Department of Signal Processing, College of Engineering, Pune, India. Her areas of interest are image processing and deep learning.

Ms. S. S. Harshini Padmanabhuni is pursuing B.Tech. and M.Tech. (dual degree) in the Computer Science and Engineering Department at National Institute of Technology Rourkela, India. Her areas of interest are machine learning, computer network, information security, and data mining.

Mr. Debachudamani Prusti is a Research Scholar and pursuing his research work in the Computer Science and Engineering Department at National Institute of Technology Rourkela, India. His areas of interest are machine learning, information security, application of data mining, and algorithm. He has received his B.Tech. degree in Information Technology from VSSUT Burla, Odisha, India and M.Tech. degree in Computer Science and Engineering from National Institute of Technology Rourkela, India.

Dr. Balwant Raj did B.Tech in ECE from Government Engineering College Bathinda, M.Tech from Beant Engineering College Gurdaspur (Government of Punjab) and PhD from GNDU Amritsar, Punjab India. He is currently working as an assistant professor in the Department of Electronics and Communication Engineering at Panjab University's Swami Sarvanand Giri Regional Centre, Hoshiarpur, Punjab, India. He is having around 18 years teaching experience. Dr. Raj has published 2 books, more than 15 research papers in reputed international journals and also he has published several national/international conference papers and book chapters in the recognized societies. His research interest includes semiconductor devices, embedded systems, VLSI Design, communication system etc.

Dr. Balwinder Raj (MIEEE'2006) did B. Tech, Electronics Engineering (PTU Jalandhar), M. Tech-Microelectronics (PU Chandigarh) and Ph.D-VLSI Design (IIT Roorkee), India in 2004, 2006 and 2010 respectively. For further research work, European Commission awarded him "Erasmus Mundus" Mobility of life research fellowship for postdoc research work at University of Rome, Tor Vergata, Italy in 2010–2011.

Dr. Raj received India4EU (India for European Commission) Fellowship and worked as visiting researcher at KTH University, Stockholm, Sweden, Oct–Nov 2013. He also visited Aalto University Finland as visiting researcher during June 2017. Currently, he is working as Associate Professor at National Institute of Technical Teachers Training and Research Chandigarh, India since Dec 2019. Earlier, he was worked at National Institute of Technology (NIT Jalandhar), Punjab, India from May 2012 to Dec 2019. Dr. Raj also worked as Assistant Professor at ABV-IIITM Gwalior (An autonomous institute established by Ministry of HRD, Govt. of India) July 2011 to April 2012. He had received Best Teacher Award from Indian Society for Technical Education (ISTE) New Delhi in 26 July 2013. Dr. Raj received Young Scientist Award from Punjab Academy of Sciences during 18th Punjab Science Congress held on 9 Feb 2015. He has also received research paper award in International Conference on Electrical and Electronics Engineering held at Pattaya, Thailand from 11 to 12 July 2015. Dr. Raj has authored/co-authored 3 books, 8 book chapters and more than 70 research papers in peer reviewed international/national journals and conferences. His areas of interest in research are Classical/Non-Classical Nanoscale Semiconductor Device Modeling; Nanoelectronic and their applications in hardware security, sensors and circuit design, FinFET based Memory design, Low Power VLSI Design, Digital/Analog VLSI Design and FPGA implementation.

Prof. Santanu Kumar Rath is a Professor in the Department of Computer Science and Engineering, National Institute of Technology Rourkela, India since 1988. His research interests are in software engineering, system engineering, machine learning, bioinformatics, and management. He has published a good number of articles in international journals and conferences in these areas. He is a senior member of the IEEE, USA, ACM, USA, and Petri Net Society, Germany.

Kaustubh Vaman Sakhare received Master and Bachelor of Engineering degrees from Sant Gajanan Maharaja College of Engineering, Shegaon, Amaravati University and Government College of Engineering, Jalgaon North Maharashtra University, India in 2007 and 2003, respectively. Currently, he is Ph.D. scholar at the Center of Excellence in Signal Processing at College of Engineering, Pune. His areas of interest are deep learning for computer vision, machine vision, pattern recognition, joint time frequency analysis, etc.

Mr. Atul Sharma is presently working as Assistant Professor in the Department of Computer Engineering, University Institute of Engineering & Technology, Kurukshetra University, Kurukshetra. He has nearly 5 years of teaching experience. He has been involved in the research related to wireless networks especially delay tolerance networks (DTNs) for last

5 years. He received his Master's degree from Deenbandhu Chhotu Ram University of Science and Technology (DCRUST), Murthal, India. His current research interests mainly include cloud computing, wireless sensor networks, and edge computing. It will decrease the number of notifications from device to the owner and make life better and easier. He has published 35 research papers in referred international journals and international conferences. He also has got the Editorial/Reviewer Board membership of more than 15 international conferences till now.

Dr. Jeetendra Singh is currently working as an assistant professor in the Department of Electronics and Communication Engineering at National Institute of Technology, Sikkim. Also, he has worked in the same position in the department of ECE at NIT Kurukshetra, Haryana, from 2012 to 2013 and at NIT Jalandhar, Punjab from 2013 to 2015. He received his B.Tech degree in Electronics and Communication Engineering from Uttar Pradesh Technical University, Lucknow, India, in 2009 and the M.Tech degree in Microwave Electronics from the University of Delhi, New Delhi, India, in 2012. He received his Ph.D. degree in VLSI Design from Dr. B. R. Ambedkar NIT Jalandhar. Dr. Singh has published more than 15 research papers in reputed international journals and also he has published several national/international conference papers and book chapters in the recognized societies. His research interest includes memristive devices, novel advanced semiconductor devices, TFETs, Analog/Digital VLSI Design, MOS Device-based sensors design, etc.

Mr. Eric Tjon is a graduate student at San Jose State University, residing in the Bay Area of California. He is currently working towards an M.S. degree in Computer Science. His research interests include machine learning, generative adversarial networks, adversarial attacks, and computer vision. Currently, he is researching on DeepFake detection methods on the DeepFake Detection Challenge dataset on Kaggle. His previous research projects include analyzing peer-to-peer loan data and exploring adversarial attacks on image classification. Mr. Tjon is also an officer in the Machine Learning Club at SJSU, serving as the President for the Spring 2020 semester.

Dr. Ashok Kumar Turuk is a Professor at National Institute of Technology Rourkela. He received his Master's degree from NIT Rourkela and Ph.D. degree from Indian Institute of Technology Kharagpur in 2000 and 2005, respectively. He has received many research grants from different government agencies. Dr. Turuk has published more than 200 research articles in the field of communication and networking. His areas of interest include optical networking, sensor networking, cloud computing, and security. He is a member of IEEE.

Prof. Shirshu Varma graduated with a degree in Electronics and Communication Engineering from Allahabad University, Allahabad, India; postgraduated with a degree in Communication Engineering from BIT Mesra, Ranchi, India; and received the Ph.D. degree in Optical Communication from the University of Lucknow, Lucknow, India. He was Lecturer, Senior Lecturer, and IT Consultant with BIT Mesra Ranchi, IET Lucknow, and C-DAC Noida. He is working as a Professor at IIIT, Allahabad. His areas of interest are wireless sensor networks, mobile computing, digital signal processing, and optical communication systems.

Dr. R. Leela Velusamy obtained her degree in Electronics and Communication Engineering in 1986 from REC Tiruchirappalli and postgraduate degree in Computer Science and Engineering in 1989 from REC Tiruchirappalli. She was awarded Ph.D. degree by the NIT Tiruchirappalli in 2010. Since 1989, she has been in teaching profession and currently she is a Professor in the Department of C.S.E, N.I.T., Tiruchirappalli, Tamil Nadu, India. Her research interests include QoS routing, ad hoc and wireless sensor networks, internet of things, social networks, and digital forensics.

Vibha Vyas received the Ph.D. degree and M.E. degree from College of Engineering, Pune (COEP) under University of Pune in 2010 and 2002, respectively. Her main work was research in the field of geometric transform invariant texture analysis. At present, she is serving as an Associate Professor at Electronics and Telecommunication Engineering Department of College of Engineering Pune (COEP), which is an autonomous institute of Government of Maharashtra, India. Her areas of interest are pattern recognition, texture analysis, signal processing, coding schemes, VLSI-embedded systems, and antenna. She is guiding many Ph.D. and M.Tech. students in the said domain.

Mr. Awaneesh Kumar Yadav is a postgraduate student in the Department of Computer Science and Engineering, NIT Rourkela. Presently he is working on design and development of authentication protocols for the next-generation cellular network. His recent research includes secure inter-living of virtual machine at the base band pool of C-RAN.

Ir. Dr. Kim Ho Yeap received his Bachelor of Engineering from Universiti Teknologi Petronas in 2004, Master of Science from Universiti Kebangsaan Malaysia in 2005, and Ph.D. from Universiti Tunku Abdul Rahman in 2011. In 2008 and 2015, respectively, he underwent research attachment at University of Oxford and Nippon Institute of Technology. Yeap is a senior member of the IEEE and a member of the IET. He is also a Chartered Engineer registered with the UK Engineering Council and a

Professional Engineer registered with the Board of Engineers Malaysia. He is currently an Associate Professor in Universiti Tunku Abdul Rahman. Throughout his career, Yeap has served in various administrative capacities, including the Head of Programme and the Head of Department of Electronic Engineering.

Yeap's research areas of interest are artificial intelligence, electromagnetics, and microelectronics. To date, he has published 72 journal papers, 39 conference proceedings, 6 books, and 12 book chapters. Yeap is the external examiner and external course assessor of Wawasan Open University. He is the Editor in Chief of the *i-manager's Journal on Digital Signal Processing*. Yeap has been given various awards which include the University Teaching Excellence award, 4 Kudos awards from Intel Microelectronics, and 19 research grants.

Chapter 1

The evolution of robotic systems

Overview and its application in modern times

Reinaldo Padilha França, Ana Carolina Borges Monteiro, Rangel Arthu, and Yuzo Iano
State University of Campinas (Unicamp)

CONTENTS

1.1 INTRODUCTION

From the beginning of the modern age, society has imagined a future supported by technology, where man and machine live and share functions both at home and at work. The terms robotics, mechatronics, computer vision, and artificial intelligence (AI) have become very popular in recent years, although the definitions are still somewhat controversial, yet they are aligned with the idea of machines that can work for humans, doing unpleasant or dangerous work, or that is, electronic employees who do not strike or ask for pay increases (Lu 2017).

In times of digital transformation and industry 4.0, industrial robotics is a promising area with high potential since this area specializes in working directly with task automation that can be seen in the automotive industry and other assembly lines, with the growing replacement of man by machine in strenuous and repetitive tasks. Using robots to perform tasks on the shop floor is called industrial robotics, which is how companies have been able to strategically apply automation, streamlining processes, reducing waste and risk of failure, and structuring uninterrupted operation much more easily. The use of robots in companies has several benefits since they are recommended for performing standardized tasks that

depend on a limited number of variables and are appropriate to perform the mass demands (Bahrin et al. 2016).

The basic idea of a practical robot is to replace humans in repetitive, dangerous, or tiring tasks. These tasks are mainly linked to the manufacture of objects on a production line, which means that the first idea of practical robots is rightly associated with the industry. So, the so-called industrial robots fall into this category as operating machines, manufacturing things like cars, appliances, and electronic equipment to replace human workers on a production line (Petersen and Rohith 2017; Kiggins 2018).

This technology has such a wide reach that it can also be used in many proposals for building smart cities in the coming decades. Computational Vision in Robotics can be seen as the approach to the development of methods, techniques, tools, and technologies for robotics coupled with computational vision, pattern recognition, and digital image processing since its applications have a special interest in environmental monitoring, enhancement of scene in images and videos, intelligent mobile robotics, and targeted navigation in recognizing behavior in images and videos (Das et al. 2019).

However, its applications are numerous from data entry, customer dialogue guidance, order entry, machine loading and unloading, packing, pressing and stamping, product life testing, boxing, adhesive application, lines, filling, assembly, and handling of products, among other industrial activities. The result is the reduction of failures and processing time of these activities, generating cost savings, so innovations in robotics were quickly installed with low implementation cost and viability of the investment (Siciliano and Khatib 2016).

Related to these studies is AI with modern robot designs and mechatronic devices including some degree of intelligence, determining this degree of intelligence existing in a machine with the kind of action it performs. With the evolution of technology, the application of AI in the industry makes the systems can be controlled from pre-established commands, adding a more human character to the machine, making it able to contribute to a decision making. A more strategic decision is where a robot identifies the time to stop packing products or pause the process for certain security holes (Hamet and Tremblay 2017; Son 2018).

The use of robotics in healthcare has increased dramatically due to the advantages of its application, both for the patient, in the clinical aspect, and in the economic parameters of this tool. Thus, less invasive surgery, reduced pain and trauma for patients, and faster recovery are observed. In addition, robotics can shorten the patient's recovery time in certain procedures, avoiding complications and ensuring hospitalization turnover, where it is noteworthy that technology has led the healthcare segment to great changes (Hamet and Tremblay 2017; Taylor et al. 2020).

Therefore, this chapter aims to provide an updated overview of Robotic Systems and its evolution and branch of application potential in the areas

of Computer Vision as well as its growing performance in Healthcare Systems, showing and approaching its success relationship, with a concise bibliographic background, categorizing and synthesizing the potential of both technologies.

1.2 ROBOTIC SYSTEMS

Initially, it was the industries that benefited most from the development of robotics by increasing the production and elimination of dangerous tasks previously performed by humans.

In robotics, the **agent** is considered a computational system situated in a given environment, which has the perception of that environment through sensors and acts autonomously in that environment through actuators to perform a given function, and the **robot** is an active artificial agent that has a body and whose environment is the physical real world, yet an autonomous **mobile robot** is a robot capable of autonomously moving around in its environment without relying on external systems, able to make its own decisions using the feedback you get from your environment (Niku 2020).

Sensors consist of devices that detect information about the robot and the environment in which it is immersed, transmitting information collected to the robot controller. Sensors are devices that are sensitive to a physical phenomenon such as temperature, electrical impedance, light, and other signals and that transmit by means of a signal to a measuring or control device informing the variation of the value of this phenomenon. They can be in the form of accelerometer, temperature sensors, brightness, potentiometer, ultrasonic distance, or even a camera (Niku 2020).

Vision systems are equipment made up of cameras and other devices that simulate human vision, producing images that are processed and provide information for the system to make a decision, and this type of technology is applied in various industrial segments, provided that the camera of a computer vision system plays a role analogous to that of the eye in the human visual system. A **digital image** is a digital signal that has a defined discrete range, mathematically an integer matrix, where each integer represents the gray tone of a discrete point in the image plane. A point in this matrix is called a pixel, where the resolution of an image is related to how many pixels in height and width an image has. The biggest challenge of robotics is obtaining autonomy, performing its actions without human intervention, and being necessary in the use of computational intelligence techniques in robotics (Daftry et al. 2016; Siciliano and Khatib 2016).

Robotics is the term for the study and use of robots, historically used in a short story entitled "Runaround" by Isaac Asimov in 1942, who wrote countless science fiction stories where robots played a prominent role. The scientist is still the author of the famous laws of robotics, which propose the existence of four laws applicable to robotics, being understood in a

purely fictional perspective: Law 0, a robot cannot hurt humanity, or even inaction allows this to happen; Law 1, the robot cannot injure a human, or even inaction allows this to happen; Law 2, the robot must obey the orders given by humans; and finally Law 3, the robot must protect its own existence (Williams 2016; Hamann 2018; Stone 2018).

Automation is a technology that makes use of electrical, mechanical, electronic, and computer systems, employing robots to control production processes, as in an industrial automotive assembly line. In this way, the robots are fully programmable, with movable arms employed in a variety of activities such as welding, painting, loading, and unloading machines. The motive of mobile robotics is to provide access to locations, increase productivity, reduce operation and maintenance costs, and improve product quality. This science encompasses mechanical, computing, and electronic technology, with lesser participation in AI theory, control, microelectronics, human factors, and production theory (Nayak and Padhye 2018).

Industrial robots are composed of computerized electronic control circuits and an articulated mechanism called as a robotic mechanism, with a manipulator, consisting of a mechanical arm, mechanical manipulator, and robotic manipulator (Riedel 2016).

Mobile robots are of course applicable in areas where vehicles exist or where automatic and mobile mechanical systems are useful, including medical services, food delivery, medicines, newspapers, window cleaning, and household vacuuming; **assistance** at airports, shopping centers, and factories; in **agriculture** applied in planting, pruning, pesticide treatment, crops, and watering; just as in the **forest** in disease treatment and forest clearing, lawn mowing, and even in garden maintenance; important use in **hazardous environments** such as inspection of nuclear power plants, piping, and high-voltage cables; seen acting in **construction** and demolition; in **underwater environments** in the inspection and maintenance of oil rigs, maintenance of transatlantic submarine cables, and the exploration of underwater resources; **military** use related to reconnaissance and combat vehicles, material handling, in the same way as Automated Guided Vehicles; used in **security** in relation to space surveillance; in **transport** with aircraft inspection and maintenance in the same way as automatic motorway driving; and also in **space** in planetary exploration and space exploration such as asteroids and space station maintenance. Nowadays there is a growing trend in **personal assistance** applied to disabled people, smart wheelchair and recently in **entertainment** as pet robots (Son 2018; Lu 2017; Hamet and Trembla 2017; Morris et al. 2018; Bechtsis et al. 2017; Katzschmann et al. 2018; Jentsch 2016; Costa et al. 2017).

Industrial Mechanical **Arm** anatomy is matched by the **joints** connected by relative movement joints; the arm is fixed to the base on one side and the handle on the other, relying on the joints attached to the **actuators**, to perform individually programed and controlled movements instructed by a control system. Together they perform movements individually endowed

with sensory ability and instructed by a control system. In direct **kinematics**, knowing the position of each joint, the position of the **terminal organ** can be obtained through cartesian coordinates (x, y, z). Considering the **inverse kinematics**, knowing the position of the robot end, the joint configuration is obtained to achieve this position, which are relevant for inverse kinematics in **mechanical arm** control systems (only joint positions) (Ghiet et al. 2016; Gao et al. 2019).

The dynamics of a **robotic arm** are related to speed; the robotic arm's ability to move from place to place in a short period of time; the measured accuracy of the error in the position of the terminal organ; the stability, corresponding to the oscillation around the desired position of the terminal organ; and the dynamic performance of the robotic arm associated with speed response considering stability and accuracy, provided the higher the stability, the lower the oscillation (Ghiet et al. 2016; Gao et al. 2019).

The **handle** is made up of several joints close to each other, allowing the orientation of the terminal organ, which is a type of **hand** or tool intended to perform specific tasks. The terminal organ is used to describe the hand or tool that is connected to the wrist, such as a welding gun, claws, suction cups, magnets or electromagnets, hooks, spoons, and paint sprayers, and some of these terminal organs are provided with sensors providing information about the objects (Chaoui et al. 2019).

Claws can also be two-fingered, consisting of a simple model with parallel or rotational movements, providing little versatility in the handling of objects as long as there is a limited opening of the fingers. The size of objects cannot exceed this aperture; even a cylindrical object gripper consists of two fingers with semicircles allowing it to hold cylindrical objects of different diameters, whereas a hinged gripper having similar shape as human hand, provides sufficient versatility for manipulating objects of irregular shapes and sizes. (Ji et al. 2018).

Joint types can be classified as prismatic or linear, with straight-line motion, consisting of two rods that slide together, or **torsional rotary** where the input and output links have the same direction as the axis of rotation of the joint. It may be **revolving rotary** where the input link has the same direction as the axis of rotation, or even **rotational rotary** where the inlet and outlet links are perpendicular to the joint rotation axis (Erkaya 2017).

Handler and handle nomenclature are based on the types of joints used in the link chain considering typical arm and wrist configurations for industrial robots. In this category, there is the **cartesian robot** which uses three linear joints, being the simplest configuration type, displacing the three joints relative to each other; the **scara robot** which has two rotary joints and one linear joint; and **cylindrical robots** having at the base a linear joint on which a rotary joint rests, still containing a third linear joint connected to the rotary joint (Apostoł et al. 2019).

Drive systems are those with the presence of actuators which are devices responsible for the movement of joints, or joints, considering the dynamic performance of the robot, which can be **electrically** powered by a direct current motor since the motor is a device that produces rotational motion based on the circulation of electric current under a magnetic field that produces an induced counter electromotive force. **Pneumatic** actuators are similar to hydraulic drives, but the difference is the use of air instead of oil. For oil **hydraulic** actuators containing components such as cylinder, oil pump, engine, valve, and oil tank, the engine being responsible for the flow of oil in the cylinder toward the piston that moves the joint is usually associated with larger power robots when compared to pneumatic and electric actuators, while still taking into account the lower accuracy compared to electric actuators (Lu 2017).

Typical capabilities of a mobile robot are diverse, but the main ones can be listed such as position estimation, speed, and direction determination capability, calculating position and height, measuring the position of the foot joints, calculating positions with respect to x, y, and w positions in the plane, measuring acceleration as well as calculating positions with respect to x, y, z, r, p, and w positions in the three-dimensional (3D) space, also have insights into obstacle detection, terrain classification, object recognition, and world representation with respect to environmental modeling and terrain map building. Still considering the ability of robot-oriented planning to stop and avoid an obstacle, planning a trajectory in the same way as a sequence of trajectories (route planning) can further enhance planning in the face of new information. It has the ability to control one or all-wheel rotation, control the engine speed, and track an object actuator (Lu 2017; Siciliano 2016; Niku 2020).

Over time, it has been observed that for small loads, it is cheaper to use an electric actuator (motor) since from a certain load value, it is more advantageous to use a hydraulic manipulator. However, the main objective of robotics is the integration of a set of sensor technology, control systems, and actuators performing environmental sensing, determining the best action options, and performing actions safely and reliably (Lu 2017; Siciliano 2016; Niku 2020; Williams 2016).

In this sense, a robot is an autonomous or semiautonomous device designed to function similar to a living entity, which includes electronic parts, mechanical parts, programs, and all that is necessary for it to achieve its purpose (Siciliano 2016; Niku 2020; Williams 2016).

1.3 COMPUTATIONAL VISION IN ROBOTICS

Computer Vision can be understood as the scientific area that seeks to describe the visible world from the data provided by one or more images reconstructing its properties as illumination form, and with the distribution of color, it is dedicated to developing theories and methods aimed at

the automatic extraction of useful information contained in images that are being captured by devices such as a camcorder, scanner, and camera in robots, among other types of sensors (Das et al. 2019).

This information obtained by the methods and techniques can be passed on or directed to a decision maker, enabling the action and navigation of mobile robots in a particular work environment. What is connected with robotics where together they use it is the "smart connection" between perception through sensors and action in the form of manipulators through mobile robots.

Computer vision systems work in the field of robotics trying to identify objects represented in digitized images provided by video cameras, thus allowing robots or devices to "see". In this sense, much evolution has been made in applications that require stereoscopic vision (with two cameras) such as identifying and locating 3D objects in the field of view (Corke 2017).

On the other hand, intelligent mobile robotics deals with the design and implementation of autonomous agents that interact in a given environment through the information provided by perception systems, or knowledge, acting in order to perform predetermined tasks, provided that the field of computational vision ranges from the extraction of visual and geometric features, through 3D reconstruction methods, to the recognition of objects and scenes linked to robotics with respect to land and air robot methodologies, small robot examinations, and robotic simulation systems (Pandey et al. 2017).

Robotics is characterized by its extreme multidisciplinary, being seen as the combination of themes inherent in mechanical and electrical engineering, control systems, and computer science, where wireless electronic sensors are integrated into devices with expert systems, with processing aid. Digital imaging still relies on the aid of AI technologies, building real-time object recognition, as it is required for robots or devices active in complex environments and requires computing power to recognize and identify objects, shapes, and people across of images, counting objects, cars, people for traffic applications, pedestrian traffic lights, and roads, among other scenarios (Niku 2020).

The integration of computer vision aims to provide data on the positioning of people in the environment according to the visual field and generate the interactive behavior of the robot since the main objective of the area is, in short, the development of autonomous systems capable of operating in a given environment and performing a variety of tasks (Corke 2017).

1.4 TECHNOLOGIES FOR ROBOTICS

Generally speaking, AI is a branch of computer science that works to build machines that mimic human intelligence with AI applications by giving machines the ability to perform tasks previously done by one person but

often faster and more quickly. Since an intelligent system can recognize the state it is in, it analyzes the possibilities of action and makes the decision on how to act according to its purpose (Hamann 2018).

Inspired by human thought that works essentially with pattern recognition by linking millions of neurons in the brain through biological neural networks, and through computation, AI can be defined as the study of intelligent agents which are basic components of an intelligent system, and having an idea behind this science, the construction of a system is inspired by these neural networks (Stone 2018).

For years, the development of robotics was associated with the application to function in industries which through AI has been instrumental in moving in open space and reproducing human gestures or even facial expressions as seen in entertainment-focused or company-oriented robots. Artificial neural networks are computational models inspired by the central nervous system of a living being, particularly the brain being able to perform pattern recognition as well as machine learning, applied to robotics, automatic target, and character recognition (Walczak 2019; Samarasinghe 2016).

Currently, there are telepresence robots that are controlled by connecting to a user and manipulated according to their movements, applied to medicine by performing extremely delicate surgeries, with great precision and without shivering, yet having remote control characteristics by a surgeon, and since with ever-increasing advances in AI, it would still be possible to implement algorithms that allow these robots to operate even more independently, saving effort and refining precision and detail. One of the active fields of application of AI is robotics; developing robots equipped with AI can help humans perform more complex and more efficient tasks (Hiyama et al. 2017; Samarasinghe 2016).

1.5 ROBOTIC SYSTEMS IN HEALTHCARE

Medicine is receiving more and more investments for nanotechnology, robotics, and software research, along with the growing trend in innovations such as Telemedicine and Wearable Devices, which are increasingly part of the routine of hospitals and clinics, contributing to improving techniques and improving care and scheduling consultations since the use of robotics in health has grown dramatically due to the advantages of its application, both in the economic parameters of this tool and for the patient, in the clinical aspect. On the other hand, less invasive surgeries, pain and trauma reduction, and faster recovery of patients are observed. Even considering that robotics can decrease the patient's recovery time in certain procedures, even in the case of change in the consulting hospital as well as avoiding complications (Taylor et al. 2020; Riek 2017).

The reality is that medicine is constantly changing, where each day new discoveries are made about adverse effects of medicines and the

pathophysiology of diseases, and often new diseases become recurrent, and even microorganisms gain resistance, in fact. In view of this, new antibiotics are launched against substances used to fight infections. In contrast, new robotic devices are created helping the doctor in the diagnosis and treatment of new pathologies. Persistent with the worldwide trend that medicine is increasingly based on clear data, reliable evidence proves the effectiveness of one treatment over another, moving away from the doubtful theories or individual experience of each physician (Riek 2017; Patino et al. 2016).

In this sense, technology has led the healthcare industry to great changes, where currently there are several techniques that allow the understanding of mammography analysis, hematological analysis, diabetes risk factors through algorithms that predict hospitalization, detection of pulmonary nodules, and Robotic automation studies to aid the patient and even surgical procedures. Research shows that robotic surgery can reduce a patient's length of stay by up to 20%. Robotics in medicine has proven to be a very successful partnership, and one of these advances comes with the help of robots during surgical procedures or in monitoring the patient's clinical parameters. Nourished with AI dealing with the area of robotics and informatics, it is focused on the creation of mechanisms capable of processing information in a more elaborate way (Xia et al. 2018; Monteiro et al. 2017, 2018a, b, 2020; Borges Monteiro et al. 2019; Yeh et al. 2017).

Robot movements are not autonomous since the technology depends entirely on the human being and only works with the existence of a qualified professional to operate it, where the robotic instruments are used as an extension of the surgeon's hand and obey his command. The movements are staggered, unlike humans, allowing access to the most sensitive areas of human body and perform millimeter movements with low error rate. Through a computer program, the robot performs very meticulous procedures, while the surgeon makes the planning of the procedure observing its execution, to adjust some points in case of any unforeseen event in the patient's clinical evolution. There will be deviations in the stipulated and planned trajectory, with higher safety, speed, and incredibly precise maneuvers (Yeh et al. 2017; Erkaya 2017).

AI is currently used in several areas besides information technology, such as medicine and health sectors. Machines and robots are already used to assist in diagnostics, surgical procedures, data processing, and treatment of various diseases, in addition to many types of research involving the participation of AI (Walczak 2019; Samarasinghe 2016; Riek 2017).

With the robotic arms, through a combination of visual, tactile, and auditory feedback, the movements of a surgeon's hands are faithfully reproduced, making the procedure orchestrated and very controlled. The use of robotic devices in surgeries began in urology, gynecology, and some types of thoracic and abdominal surgeries. Currently, the technique is quite

common in laparoscopic gallbladder removal procedures (Yeh et al. 2017; Ji et al. 2018; Gao et al. 2019; Ghiet et al. 2016).

The surgeon feels as if he is present at the surgery, and may also feel the touch and the firmness of the tissues that the robot arms are manipulating. Some robot technologies have a camera attached to the robot to stay still and reproduce high-definition 3D images, which are seen more clearly, facilitating the most delicate procedures and reducing the chances of damage (Jain et al. 2017; Grespan et al. 2019a; El-Ghobashy et al. 2019).

In addition to amplifying the surgeon's potential and eliminating tremor, the robots have unique skills which are the wrist rotating on seven different axes, 360 degrees. This staggering movement is a great advantage, resulting in more delicate, accurate, and precise surgery. (El-Ghobashy et al. 2019; Taylor et al. 2020).

Even the most experienced doctors may shiver a little in more difficult and difficult to access surgeries, but this does not happen in the presence of a robot, as a higher level of accuracy is easily achieved, eliminating any kind of tremor, and the doctor can still count on moving the axes 360 degrees, reaching angles that a doctor's hands would reach with great difficulty. Also, avoiding the nervousness of novice professionals facing the complexity of the clinical case in question, and in more complex surgeries, robotic devices lead to less invasive procedures and faster recovery because the surgical incision is defined for the specific region, i.e., smaller, which significantly facilitates the performance of the surgery (Grespan et al. 2019b).

Using this feature, it is possible to perform physiological procedures in very sensitive areas, such as the eyeball and large organs that demand precision, such as the heart, avoiding significant bleeding; thus, with the doctor's supervision, it has the ability to perform the procedures. With robotic equipment, it is possible to design personalized procedures with the medical team, with fewer interventions and shorter patient recovery time. Many interventions related to, for example, the uterus, digestive tract, prostate glands, heart, and brain have been performed using robots, achieving great success and excellent levels of efficiency (El-Ghobashy et al. 2018; Shepherd and Nobbenhuis 2018).

Telesurgery is the combination of two technological resources, namely, telemedicine and robotics, which enables broadening the range of monitoring of vital patient data and clinical services in both the therapeutic aspects. Telesurgery robotics represent a surgeon operating in the next room or even miles away using videoconferencing technologies (Choi et al. 2018).

Telemedicine makes it possible to diagnose diseases by quickly determining laboratory parameters or instituting robotics for high-sensitivity procedures in patients using the technology for therapeutic, diagnostic, and follow-up dimensions in the same sense as the follow-up of clinical evolution in real time can significantly modify the therapeutic conduct of the physician (Mehrotra et al. 2016).

Not only face-to-face surgeries can be performed with a robot but also those at a distance, especially in locations that do not have a surgeon specialized in the area you need, yet considering the feasibility of four-hand procedures with the specialist in the field operating remotely and the local surgeon following up on the spot, with the possibility of intervening in any situation of need. In more remote places, even if a surgeon does not have much experience in particular surgery, he can be assisted by an expert with telemedicine systems, avoiding unnecessary travel of critical patients to more advanced locations ensuring less impact on the patient (Fekri et al. 2018; Hung et al. 2018).

Through robotics, the surgeon follows the surgery interfering at any time, supporting the local surgeon, provided that everything is previously studied considering all the risks that compromise the patient's life. Thus, through a robotic surgery system, productivity and efficiency of interventions are increased, where no surgery will be impossible because of isolated sites (Choi et al. 2018; Grespan et al. 2019a).

Telepresence is another possibility of robotics in health services, which has robots installed in various sectors, promoting quick access to medical specialists showing the results of tests already performed on the patient and discussing cases with the support team, facilitating decision making, and enabling communication with the doctor discussing less urgent interventions since these telepresence robots can operate for up to 80 h, being 20 h of intermittent use, and can move easily, having collision avoidance sensors, with a network connection such as location Wi-Fi and controlled by mobile devices like tablets or smartphones (Koceski and Koceska 2016).

The use of robotics is also present in nursing since the workload of the nursing staff is high requiring attention in patient care, both regarding the administration of medications as well as regarding the procedures as the monitoring of vital parameters. In this sense, the nursing robots facilitate some activities, helping nurses to deal with more complex situations and minimizing repetitive actions since robotics works to measure clinical data, improving the process of drug administration and storing via electronic medical records. (Locsin and Ito 2018).

In this context, the benefits of using robotics in healthcare provide clinical and economic advantages since it is possible to count on the support of specialists by videoconferencing in the same way as the greater precision in procedures facilitating the performance of remote surgeries. Previously this efficiency brought by robotics was related to the type of automation of processes previously performed repeatedly by healthcare professionals reducing preventable errors and increasing productivity, which are crucial factors in a health facility (Jain et al. 2017; Grespan et al. 2019b; El-Ghobashy et al. 2019; Mehrotra et al. 2016).

Robotics in medicine shortens the length of stay, with less bleeding, little surgical aggression, and less trauma, and consequently, reduces the risk of

infection, as large cuts, deep scars, and severe bleeding are avoided, thus shortening the postoperative period. However, robotics faces some barriers ahead with regard to people's mistrust, along with the institutions' resistance to the return on investments in machines and the training of people. Professionals would need to undergo a lot of updating to be able to keep up with the technological advances, which makes medical treatment more expensive. But even so, the gains of robotics in medicine are countless simply because we have a digital image of surgery for a magnifying glass camera, while still considering the gains in intraoperative decision making that need to be quick and assertive (Koceski and Koceska 2016; Locsin and Ito 2018).

1.6 SCIENTIFIC REVIEW

In 2016, reliability assessment models of robotics software implemented based on artificial neural network methods were researched since its implementation developed a special type of a vertical layer neural network, increasing the accuracy of the model leading to the accumulation of layers in the network neural. The research was dedicated to the design of the complex reliability of different function control systems software based on the research related to the adaptation of the procedure-oriented model to the object-oriented software, its reduction of a known formula for the reliability model prediction. For creation of an artificial neural network, it was once created to predict and evaluate the complex reliability of the software, as well as the description of a new embodiment of the constructed neural network (Sheptunov et al. 2016).

In 2017, research was conducted on sparkling wines and carbonated drinks from the perspective of computer vision as a technique applied in the food industry helping in quality control and process. In this sense, a robotic leaker (FIZZeyeRobot) capable of normalizing the variability of foam and bubble development during filling was presented since a camera was attached to capture video for evaluation of various foam quality parameters, including the foam's ability to form, drainage capacity, and bubble count and allometry. The highest quality scores were positively correlated with foam stability and negatively correlated with foam dissipation velocity and collar height, as shown in the results. Thus, the wines were grouped according to their respective bubble and foam characteristics considering chemical parameters and quality indices. The developed technique objectively evaluated the characteristics of sparkling wine foam using image analysis, maintaining a fast, repeatable, economical, and reliable robotic method, and also taking into account that the parameters obtained have the potential to help unveil contributing factors of wine quality (Condé et al. 2017).

In 2018, it was studied that one of the biggest challenges of a real-time autonomous robotic system for monitoring a building is the mapping, location, and simultaneous navigation over its lifetime with little or no human intervention over the active research context of computer vision and robotics. In a monocular vision system, the understanding of a real-time scene is computationally heavy considering state-of-the-art algorithms tested on robust desktops and servers with high CPU and GPU capabilities that affect deployment and mobility for real-world applications. In this scenario, an integrated robotic computer vision system was proposed generating a real-world spatial map with respect to obstacles and traversable space in the environment through integration and contextual awareness with visual Simultaneously Localization and Mapping (SLAM) present in a solo robotic agent. Research has demonstrated hardware utilization and performance for three different outdoor environments representing the applicability of this pipeline to various near real-time outdoor scenes, considering that the system is standalone and does not require user input, demonstrating the potential of a computer vision system for autonomous navigation (Asadi et al. 2018).

In 2019, the role of assisted robotic surgery in the treatment of contemporary urolithiasis in its infancy was studied, since, in rare clinical circumstances, such as the presence of large ureter and pelvic stones, endourological techniques are not appropriate. In this context, robotic-assisted laparoscopic ureterolithotomy, robotic-assisted laparoscopic pyelolithotomy with or without simultaneous pyeloplasty, and robotic-assisted laparoscopic nephrolithotomy are described for performing complex calculations, taking into account the technical challenges that flexible ureteroscopy has related to the development of ureteroscopy, assisted by robotics (Suntharasivam et al. 2019).

In 2020, it was seen that experienced surgeons studied the musculoskeletal workload during laparoscopic surgery compared with robotic-assisted laparoscopy (RALS). This study was based on the proportion that 70%–90% of surgeons who perform laparoscopic surgery report musculoskeletal symptoms with a higher incidence on the neck and shoulders, and are counterpoised with data on the potential ergonomic benefits of RALS in a clinical setting, but are limited. Thus, 12 experienced surgeons in laparoscopic surgery and RALS were engaged in the process, performing two hysterectomies on the same day. Laparoscopic surgery was performed standing, and RALS performed sitting, with the support of the forearm and head. The bipolar surface electromyogram was recorded in several muscles during maximal contractions, the intervals per minute plus static muscle activation were calculated, and the perceived exertion before and immediately after each surgery was also evaluated. Moreover, these results showed that RALS was significantly less physically demanding compared to laparoscopic surgery being considered less arduous for surgeons. However, for

both types of surgeries, there is still room for improvement in working conditions (Dalsgaard et al. 2020).

1.7 APPLICATIONS OF ROBOTICS

There are many uses of robots today in health since the market today is much more promising, considering that robot expansion in many areas of practice could help save lives, prevent contagion, increase the accuracy of doctors' movements, and ensure access to complex procedures at hostile or remote locations. Considering why robots are becoming commonplace is a matter of the economics of automation since automation means that a process can move on its own and work alone with little or no direct control from a human. Automation brings economic advantages to the workplace, thus increasing productivity. One key economic advantage is that there is less human involvement, which means that a costly expense, that is, of human labor, is reduced, thereby reducing human capital by increasing spending on automation equipment (Lu 2017).

Repetitive activities, collection of blood tests, and obtaining blood pressure, among others, are examples of the use of robotics in care, which can now be carried out by robot nurses, leaving nursing more focused on treatment and relationship with the patient. Through augmented reality, the robot can see exactly where to place the needle and accurately and assertively insert the needle into the patient's vein (Monteiro et al. 2018a, b; Locsin and Ito 2018; Koceski and Koceska 2016).

In many emergency situations, robots can be used for immediate treatment of burns, heart attacks, and many other conditions that would require medical procedures, even considering disasters such as rescues from the wreckage from still glowing accidents or earthquakes, vehicular accidents, and many others (Koceski and Koceska 2016).

In the industry, robots have been in business for a long time since robotization is still going to grow more in the coming years, considering that today the automotive industry is the largest consumer of robots worldwide. From the historical point of view, one of the main concerns regarding the use of robots in the industry is the reduction of job vacancies by replacing people. They can be used in automated transportation and cargo transportation vehicles on a production line, placed to perform risky human functions such as transporting glass for car assembly, providing increased productivity, better utilization of raw materials, cost savings, energy saving, and the ability to assemble miniature parts and coordinate complex movements (Lu 2017).

The automation rate in the automotive industry is over 90%, which in some stages of production can reach 100%, and the automotive industry is one of the fields that uses robotics technology the most, with its robots scheduled to replace each time human labor, and withstand stressful, repetitive, or hazardous working conditions (Bechtsis et al. 2017).

Contrary to popular myth at the time, the impact on the market was positive, including creating jobs in previously unimagined areas since technicians are required to program the robot and technicians to maintain it, relating more to the quality issue and safety since the robot does the grossest and most repetitive service, resulting in increased opportunities that need more qualification (Siciliano 2016).

Military air and ground monitoring systems include the use of robots in the case of increasing aerial robotics not only in aerospace engineering but for military applications, expressing a growing market for such applications in the coming years where robot aircraft are capable of transmitting live images of areas of interest, enabling a new dynamics of command actions, with greater intelligence gain and faster action-taking (Jentsch 2016).

In the health field, as described in previous sections, robots are present in various forms, including support for the elderly and disabled, with automated prostheses and wheelchairs since patients with paralysis can walk again through the use of exoskeletons. Equivalently for people who have lost part of their movements due to stroke, weakened limbs due to old age or other causes, even considering the stabilization of a patient suffering from Parkinson's disease. The application of robots in medicine allows remote surgical interventions to be performed through the teleoperation of robots and can be used in delicate surgeries that require maximum movement precision. In surgical procedures such as video endoscopy that use robots as surgeons are guided by the remote control or voice commands, thus replacing doctors in critical surgeries that were never imagined before in surgeries related to stomach, heart, arteries, kidneys and even brain (Taylor et al. 2020).

The ability to perform tasks with far greater precision and quality than manual labor is one of the key factors linked to the continuous and high production speed of robots, automated processes are not considered to get slow down due to the external physical factors but consider the control over how work can be made scalable (Bahrin et al. 2016).

Robots often perform household chores such as the trend of robotic vacuum cleaners and robotic lawnmowers, and they can even sweep the floor, clean glass and pools, and take care of the garden since research in home automation aims at developing a society in which robots coexists with humans and performs some tasks making human lives easier (Thompson et al. 2018).

Robots still appear as a playful and active form of learning, and can be applied in elementary and high school education, still being applied in participation in the tournament, which makes young people's contact with technology encouraging the creation of more connected future professionals. It also helps children to learn physics and mathematics concepts in a simpler way and also useful in making the concepts more understandable for higher education studies (Miller and Nourbakhsh 2016).

Robots can also be used in logistics and chemical testing where they can perform the task of bringing sensitive drugs, can transfer the products from one place to another with less damage. When it comes to the risk of contact with a particular material, robotics is useful in chemical testing or clothing testing to protect humans from hazardous procedures and environments with infectious contagious epidemics (Wahrmann et al. 2019).

A growing trend in modern times is the use of robotics in psychiatric treatments. There are a number of robots used for relieving from the feeling of loneliness and treating people with mental disorders, by acting as companions in addition to teleconsultations and remote monitoring of these patients by psychologists and psychiatrists (Riek 2016).

1.8 FUTURE TRENDS

The future of medical imaging and PACS software is directly related to the changes brought about by robotics since large resonance, tomography, and the like will be replaced by nanorobots that make it easier to capture much more accurate images being interpreted by experts. Considering the advent of AI, these images will be interpreted so that the report can be completed faster and more productively than is currently done, allowing a more immediate treatment of the health problems presented by the patients, and thus reducing complications and inherent costs.

In the industrial field, even the most advanced robots will need the presence of increasingly specialized technicians for their maintenance and programing, so still there is a need of possible development of robots capable of self-monitoring and acting on themselves, which is a great challenge in the digital world.

In the medical field, there will be an even greater increase in minimally invasive surgeries, and in many cases, brain surgery that seems to be infeasible are becoming possible.

In the nanorobotic area, the future promises are bright. Nanorobots that are inserted into capsules and can be swallowed by patients are being developed. Once dissolved in the human body, they release from the capsules and, with the help of specialized technicians and doctors, attack and destroy bacteria and even cure cauterizing wounds, as well as involve in many other activities such as clearing fat walls of the heart arteries, thus eliminating the danger of heart attacks.

1.9 CONCLUSIONS

The presence of robotics is increasing in our reality. Main innovations of robotics in medicine are transforming medical consultations and treatments, which were previously restricted to laboratory consultations

and evaluations. Combining technology with processes and diagnostics, the results for all areas are optimized. Robots are really promoting a social revolution. In healthcare, with the support of robots, doctors and nurses can optimize their work and ensure to provide more effective treatments.

AI has found its applications in almost every domain of working ranging from automated cars, healthcare to stock markets. Through the growth of technology, information production, and computer processing capabilities, AI will become much more powerful in the future, thus enabling specialized technological discoveries in robotics.

In this sense, the importance of robotics in society increases due to the increasing range of their applications that facilitate the daily life of humans, such as autonomous robots linked with AI, which has the ability to learn similarly as a human being through machine learning, and perform tasks and functions faster, more accurately, and more safely compared to humans.

REFERENCES

Apostoł, M., Kaczmarczyk, G., & Tkaczyk, K. (2019, May). SCARA robot control based on Raspberry Pi. In *2019 20th International Carpathian Control Conference (ICCC)* (pp. 1–6). IEEE.

Asadi, K., Ramshankar, H., Pullagurla, H., Bhandare, A., Shanbhag, S., Mehta, P., ... & Wu, T. (2018). Building an integrated mobile robotic system for real-time applications in construction. arXiv preprint arXiv:1803.01745.

Bahrin, M. A. K., Othman, M. F., Azli, N. N., & Talib, M. F. (2016). Industry 4.0: A review on industrial automation and robotic. *Jurnal Teknologi*, 78(6–13), 137–143.

Bechtsis, D., Tsolakis, N., Vlachos, D., & Iakovou, E. (2017). Sustainable supply chain management in the digitalisation era: The impact of automated guided vehicles. *Journal of Cleaner Production*, 142, 3970–3984.

Chaoui, M. D., Léonard, F., & Abba, G. (2019). Improving Surface Roughness in Robotic Grinding Process. In *ROMANSY 22–Robot Design, Dynamics and Control* (pp. 363–369). Springer, Cham.

Choi, P. J., Oskouian, R. J., & Tubbs, R. S. (2018). Telesurgery: past, present, and future. *Cureus*, 10(5).

Condé, B. C., Fuentes, S., Caron, M., Xiao, D., Collmann, R., & Howell, K. S. (2017). Development of a robotic and computer vision method to assess foam quality in sparkling wines. *Food Control*, 71, 383–392.

Corke, P. (2017). *Robotics, Vision and Control: Fundamental Algorithms in MATLAB® Second, Completely Revised* (Vol. 118). Springer, New York.

Costa, A., Julian, V., & Novais, P. (Eds.). (2017). *Personal Assistants: Emerging Computational Technologies* (Vol. 132). Springer, New York.

Daftry, S., Zeng, S., Bagnell, J. A., & Hebert, M. (2016, October). Introspective perception: learning to predict failures in vision systems. In *2016 IEEE/RSJ International Conference on Intelligent Robots and Systems (IROS)* (pp. 1743–1750). IEEE.

Dalsgaard, T., Jensen, M. D., Hartwell, D., Mosgaard, B. J., Jørgensen, A., & Jensen, B. R. (2020). Robotic surgery is less physically demanding than laparoscopic surgery: Paired cross-sectional study. *Annals of Surgery*, 271(1), 106–113.

Das, S. K., Dash, S., & Rout, B. K. (2019). An integrative approach for tracking of mobile robot with vision sensor. *International Journal of Computational Vision and Robotics*, 9(2), 111–131.

El-Ghobashy, A., Ind, T., Persson, J., & Magrina, J. F. (Eds.). (2018). *Textbook of Gynecologic Robotic Surgery*. Springer, Cham.

Erkaya, S. (2017). Effects of joint clearance on motion accuracy of robotic manipulators. *Journal of Mechanical Engineering*, 64(2).

Fekri, P., Setoodeh, P., Khosravian, F., Safavi, A. A., & Zadeh, M. H. (2018). Towards Deep Secure Tele-Surgery. In *Proceedings of the International Conference on Scientific Computing (CSC)* (pp. 81–86). The Steering Committee of The World Congress in Computer Science, Computer Engineering and Applied Computing (WorldComp).

Gao, J., Chen, Y., & Li, F. (2019). Kinect-based motion recognition tracking robotic arm platform. *Intelligent Control and Automation*, 10(03), 79.

Ghiet, A. M. A., Dakhil, M. A., Dayab, A. K., Ali Saeid, A., & Abdellatif, B. A. B. A. (2016). Design and development of robotic arm for cutting trees [R]. *ResearchGate*. December, P1.

Grespan, L., Fiorini, P., & Colucci, G. (2019a). Looking Ahead: The Future of Robotic Surgery. In *The Route to Patient Safety in Robotic Surgery* (pp. 157–162). Springer, Cham.

Grespan, L., Fiorini, P., & Colucci, G. (2019b). Measuring Safety in Robotic Surgery. In *The Route to Patient Safety in Robotic Surgery* (pp. 37–41). Springer, Cham.

Hamann, H. (2018). Short Introduction to Robotics. In *Swarm Robotics: A Formal Approach* (pp. 33–55). Springer, Cham.

Hamet, P., & Tremblay, J. (2017). Artificial intelligence in medicine. *Metabolism*, 69, S36–S40.

Hiyama, A., Kosugi, A., Fukuda, K., Kobayashi, M., & Hirose, M. (2017, July). Facilitating Remote Communication Between Senior Communities with Telepresence Robots. In *International Conference on Human Aspects of IT for the Aged Population* (pp. 501–515). Springer, Cham.

Hung, A. J., Chen, J., Shah, A., & Gill, I. S. (2018). Telementoring and telesurgery for minimally invasive procedures. *The Journal of Urology*, 199(2), 355–369.

Jain, K., Weinstein, G. S., O'Malley, B. W., & Newman, J. G. (2017). Robotic Surgery Training. In *Atlas of Head and Neck Robotic Surgery* (pp. 27–31). Springer, Cham.

Jentsch, F. (2016). *Human-Robot Interactions in Future Military Operations*. CRC Press, Boca Raton.

Ji, A., Zhao, Z., Manoonpong, P., Wang, W., Chen, G., & Dai, Z. (2018). A bio-inspired climbing robot with flexible pads and claws. *Journal of Bionic Engineering*, 15(2), 368–378.

Katzschmann, R. K., DelPreto, J., MacCurdy, R., & Rus, D. (2018). Exploration of underwater life with an acoustically controlled soft robotic fish. *Science Robotics*, 3(16), eaar3449.

Kiggins, R. (2018). Robots and Political Economy. In *The Political Economy of Robots* (pp. 1–16). Palgrave Macmillan, Cham.

Koceski, S., & Koceska, N. (2016). Evaluation of an assistive telepresence robot for elderly healthcare. *Journal of Medical Systems*, 40(5), 121.

Locsin, R. C., & Ito, H. (2018). Can humanoid nurse robots replace human nurses? *Journal of Nursing*, 5(1), 1.

Lu, Y. (2017). Industry 4.0: A survey on technologies, applications and open research issues. *Journal of Industrial Information Integration*, 6, 1–10.

Mehrotra, A., Jena, A. B., Busch, A. B., Souza, J., Uscher-Pines, L., & Landon, B. E. (2016). Utilization of telemedicine among rural Medicare beneficiaries. *JAMA*, 315(18), 2015–2016.

Miller, D. P., & Nourbakhsh, I. (2016). Robotics for Education. In *Springer Handbook of Robotics* (pp. 2115–2134). Springer, Cham.

Monteiro, A. C. B., Iano, Y., & França, R. P. (2017). An improved and fast methodology for automatic detecting and counting of red and white blood cells using watershed transform. *VIII Simpósio de Instrumentação e Imagens Médicas (SIIM)/VII Simpósio de Processamento de Sinais da UNICAMP.*

Monteiro, A. C. B., Iano, Y., França, R. P., & Arthur, R. (2019). Medical-laboratory algorithm WTH-MO for segmentation of digital images of blood cells: a new methodology for making hemograms. *Int J Simul Syst Sci Technol*, 20(Suppl 1), 19.1–19.5.

Monteiro, A. C. B., Iano, Y., Fraça, R. P., & Arthur, R. (2020). Development of a laboratory medical algorithm for simultaneous detection and counting of erythrocytes and leukocytes in digital images of a blood smear. Deep Learning Techniques for Biomedical and Health Informatics, 165.

Monteiro, A. C. B., Iano, Y., França, R. P., & Arthur, R. (2018a, October). Methodology of High Accuracy, Sensitivity and Specificity in the Counts of Erythrocytes and Leukocytes in Blood Smear Images. In *Brazilian Technology Symposium* (pp. 79–90). Springer, Cham.

Monteiro, A. C. B., Iano, Y., França, R. P., Arthur, R., & Estrela, V. V. (2018b, October). A Comparative Study Between Methodologies Based on the Hough Transform and Watershed Transform on the Blood Cell Count. In *Brazilian Technology Symposium* (pp. 65–78). Springer, Cham.

Morris, K. J., Samonin, V., Anderson, J., Lau, M. C., & Baltes, J. (2018, June). Robot Magic: A Robust Interactive Humanoid Entertainment Robot. In *International Conference on Industrial, Engineering and Other Applications of Applied Intelligent Systems* (pp. 245–256). Springer, Cham.

Nayak, R., & Padhye, R. (2018). Introduction to Automation in Garment Manufacturing. In *Automation in Garment Manufacturing* (pp. 1–27). Woodhead Publishing, Cambridge.

Niku, S. B. (2020). *Introduction to Robotics: Analysis, Control, Applications.* John Wiley & Sons, Hoboken, NJ.

Pandey, A., Pandey, S., & Parhi, D. R. (2017). Mobile robot navigation and obstacle avoidance techniques: A review. *International Journal of Automotive Technology*, 2(3), 96–105.

Patino, T., Mestre, R., & Sanchez, S. (2016). Miniaturized soft bio-hybrid robotics: A step forward into healthcare applications. *Lab on a Chip*, 16(19), 3626–3630.

Petersen, B. L., & Rohith, G. P. (2017). *How Robotic Process Automation and Artificial Intelligence Will Change Outsourcing.* Mayer Brown, Brussels.

Riedel, M. (2017). U.S. Patent No. 9,821,456. Washington, DC: U.S. Patent and Trademark Office.

Riek, L. D. (2016). Robotics Technology in Mental Health Care. In *Artificial Intelligence in Behavioral and Mental Health Care* (pp. 185–203). Academic Press, San Diego, CA.

Riek, L. D. (2017). Healthcare robotics. arXiv preprint arXiv:1704.03931.

Samarasinghe, S. (2016). *Neural Networks for Applied Sciences and Engineering: From Fundamentals to Complex Pattern Recognition*. Auerbach Publications, Boca Raton, FL.

Shepherd, J. H., & Nobbenhuis, M. (2018). The Development of Robotic Surgery: Evolution or Revolution? In *Textbook of Gynecologic Robotic Surgery* (pp. 1–4). Springer, Cham.

Sheptunov, S. A., Larionov, M. V., Suhanova, N. V., Salakhov, M. R., & Solomentsev, Y. M. (2016, October). Simulating reliability of the robotic system software on the basis of artificial intelligence. In *2016 IEEE Conference on Quality Management, Transport and Information Security, Information Technologies (IT&MQ&IS)* (pp. 193–197). IEEE.

Siciliano, B., & Khatib, O. (Eds.). (2016). *Springer Handbook of Robotics*. Springer, Berlin, Heidelberg.

Son, H. I. (2018, October). Robotics and artificial intelligence: extension to agriculture and life science. In 2018 년 작물보호분야 공동 국제학술대회 (pp. 418–418).

Stone, W. L. (2018). The History of Robotics. In *Robotics and Automation Handbook* (pp. 8–19). CRC Press.

Suntharasivam, T., Mukherjee, A., Luk, A., Aboumarzouk, O., Somani, B., & Rai, B. P. (2019). The role of robotic surgery in the management of renal tract calculi. Translational Andrology and Urology.

Taylor, R. H., Kazanzides, P., Fischer, G. S., & Simaan, N. (2020). Medical Robotics and Computer-Integrated Interventional Medicine. In *Biomedical Information Technology* (pp. 617–672). Academic Press.

Thompson, K. A., Paton, S., Pottage, T., & Bennett, A. M. (2018). Sampling and inactivation of wet disseminated spores from flooring materials, using commercially available robotic vacuum cleaners. *Journal of Applied Microbiology*, 125(4), 1030–1039.

Wahrmann, D., Hildebrandt, A. C., Schuetz, C., Wittmann, R., & Rixen, D. (2019). An Autonomous and flexible robotic framework for logistics applications. *Journal of Intelligent & Robotic Systems*, 93(3–4), 419–431.

Walczak, S. (2019). Artificial Neural Networks. In *Advanced Methodologies and Technologies in Artificial Intelligence, Computer Simulation, and Human-Computer Interaction* (pp. 40–53). IGI Global.

Williams, B. (2016). *An Introduction to Robotics. Mechanics and Control of Robotics Manipulators*, Dr. Bob Productions, unpublished.

Xia, Y., Cai, W., Yang, X., & Wang, S. (2018). Computation methods for biomedical information analysis. *Journal of Healthcare Engineering*, 2018, 2.

Yeh, C. C., Spaggiari, M., Tzvetanov, I., & Oberholzer, J. (2017). Robotic pancreas transplantation in a type 1 diabetic patient with morbid obesity: A case report. *Medicine*, 96(6).

Chapter 2

Development of a humanoid robot's head with facial detection and recognition features using artificial intelligence

Kim Ho Yeap, Jiun Hao Koh, and Siu Hong Loh
Universiti Tunku Abdul Rahman Kampar Campus Jalan Universiti

Veerendra Dakulagi
Guru Nanak Dev Engineering College

CONTENTS

2.1 INTRODUCTION

Robots are no longer strangers to mankind today. With the world marching inexorably towards the fourth industrial revolution (IR 4.0), human lives are now closely intertwined with the existence of robots. Wherever we are, whatever we are doing, one may just come across one of these machines in some ways or another. The variety of robots is quite diverse, and it could range widely from simple consumer products such as the autonomous

vacuum cleaners and dishwashers to unmanned aerial vehicles (UAVs) to those bulky contraptions used in factories.

Just what exactly is a robot? This is one of those questions which are seemingly easy to figure out. But one may be surprised at how complicated and vague the real answer is when one delves deep into it. The Institute of Electrical and Electronics Engineers (IEEE) provides the following definition for robots – one that is neither too general nor specific (IEEE 2019):

> A robot is an autonomous machine capable of sensing its environment, carrying out computations to make decisions, and performing actions in the real world.

This definition succinctly shows that any machines which could sense, compute, and act can be classified as a robot. To put it more explicitly, a robot is simply a machine equipped with sensors, a microcontroller (i.e. the microchip which performs computations), and probably some transducers which convert the output voltage into some other forms of energy. It is usually built with the intention of (Ho 2004)

i. emulating a human's behavior,
ii. mitigating a human's burden, or
iii. accomplishing tasks which may seem insurmountable for humans themselves to undertake.

An android is a humanoid robot which possesses the outer appearance and characteristics of a human. When designing an android, its behaviors, such as the idiosyncrasies, gestures, articulation, and movements, have to mimic closely those of a real human being. The world itself is analogue in nature, and all living creatures in general have been used to react to this kind of environment. Robots, on the other hand, view its environment in an utterly different manner. They could only interpret stimuli in digital forms. Hence, an analogue to the digital converter is necessary to allow a robot to receive signals from its surroundings and to react to it. Because of this reason, building an android which could behave exactly the same way as a human being is virtually impossible – at least for now or the next few years to come. One of the most challenging tasks in building an android is the design and construction of its head. The head can be regarded as the most crucial part of the entire robot, since that is where the microcontroller (i.e. the 'brain' of the robot) is stored. Artificial intelligence or AI is programmed into the microcontroller, so that it could dictate how the android behaves and synchronize the movements of its body.

This chapter provides a detailed elaboration on the development of a simple android's head. The purpose of the chapter is to give readers – robotics hobbyists, in particular – a basic idea of how a robot head could be

constructed from off-the-shelf components. The construction of the robot head is rather laborious, and it involves diverse engineering fields – ranging from the mechanical construction of its frame, the design and development of the electronic circuitries, and configuring the AI algorithms and downloading then into the microcontroller. Besides the microcontroller, the robot head described in this chapter is equipped with a servomotor, passive infrared or PIR-based motion sensors, a vision sensor (i.e. a camera) and facial detection and recognition algorithms. It is capable of performing the following functions:

i. It will first detect the presence of humans in its vicinity via the PIR sensors.
ii. Once it has locked to a fixed target, the servomotor mounted to the base of the head will then rotate, orientating the position of the head to face its target.
iii. With the aid of the vision camera, the robot head will then capture the image of the target in a video stream. Upon running through an AI algorithm, the robot can then detect and recognize the countenance of the target. If the countenance matches one of those in the dataset, the details of the target will then be displayed.

2.2 CONSTRUCTION OF THE FRAME

Generally, the construction of the android's head can be divided into three stages. The first stage involves the mechanical development of the frame. This is to say that the frame has to be sufficiently spacious so that it could house the sensors, servomotor, microcontroller, and the supporting electronic circuitries. Also, the frame has to be robust to support the overall weight. Figures 2.1 and 2.2 show, respectively, the frame of the head before and after it was painted, whereas Figure 2.3 depicts the final outlook of the android's head after the electronic modules were installed into it. The frame was built mostly from ethylene-vinyl acetate (EVA) foams. The appearance

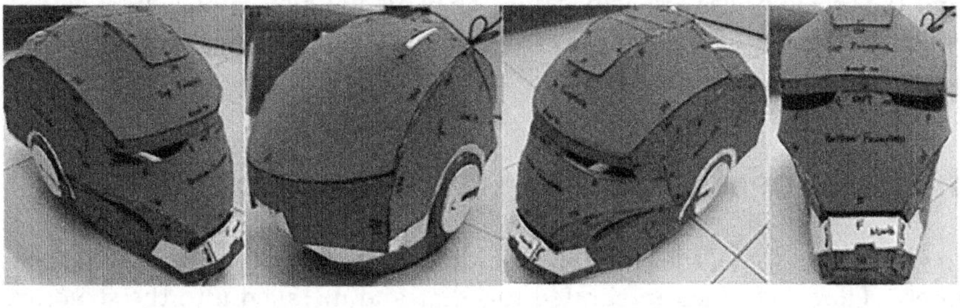

Figure 2.1 The android's head before it was sprayed.

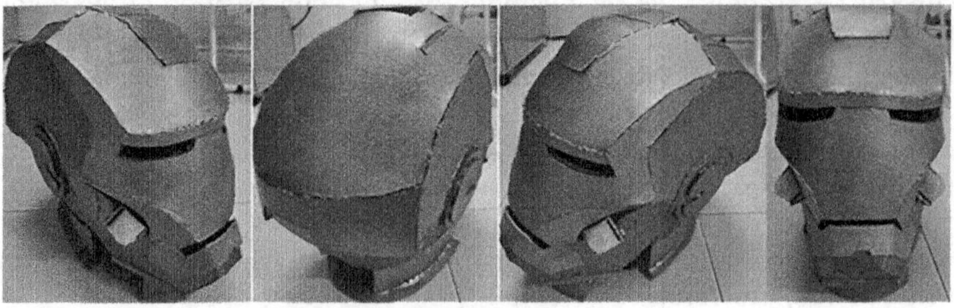

Figure 2.2 The android's head after it was sprayed.

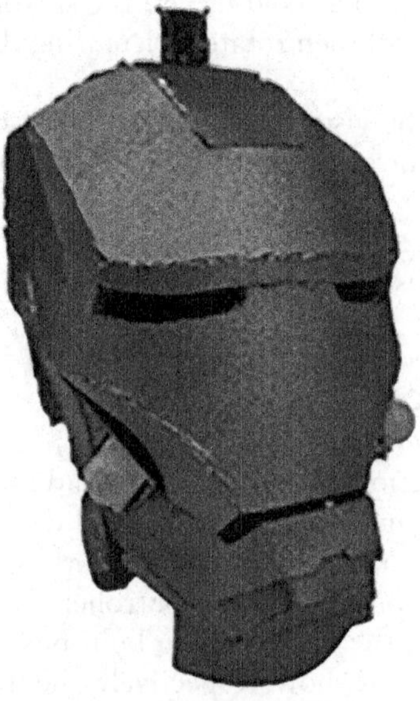

Figure 2.3 Final prototype of the android's head.

of the head is adopted from that in Instructables (JoshuaGuess 2016). EVA foams were selected because they are tough, durable, and water resistant and that their prices are affordable. They also exhibit rubber-like characteristics in terms of elasticity and flexibility. The material is therefore reminiscent of a real human skin.

2.3 DEVELOPMENT OF THE ELECTRONIC MODULES

In the second stage, the electronic modules were first built and tested separately. They were then integrated together and installed into the structure of the android's head. Further verifications were then carried out to ensure

that the system could operate properly as a whole and that the microcontroller was able to synchronize all the electronic modules. This section gives a detailed explanation on each electronic module.

2.3.1 PIR sensor module

Unlike the conventional IR sensors which work based on the detection of the reflection signal, the working principle of the PIR sensor is based on the differential change of the IR signal. In a nutshell, a PIR sensor consists of two slots. When the sensor is idle, both slots detect the same amount of IR signals. The output pin of the module stays at 0 V. When an object moves past the sensor, however, one of the slots will first be intercepted. The signal detected by both slots will therefore differ (i.e. one of the slots detects more radiation than the other). This results in a positive differential input to the comparator which subsequently triggers a positive output pulse. On the contrary, when the object leaves the sensor, the second slot will first sense the change. A negative differential input is therefore fed to the comparator, triggering a negative output pulse from the comparator. This allows the PIR sensor to distinguish the direction the living object is moving.

Figure 2.4 shows the pin layout of the PIR sensor module used in this project. As can be seen in the figure, two variable resistors, i.e. Ch1 and RL2 are used to control the operation of the sensor. Resistor Ch1 controls

Figure 2.4 A PIR sensor module.

the delay time to activate the sensor, whereas RL2 is used to adjust the detection range from 3 to 7 m. In order to test the functionality of the circuit, the 3-pin header at the bottom of the module is to be connected to the circuit. The leftmost pin is to be connected to a 5 V voltage source, whereas the rightmost pin is to be grounded. The pin at the center of the header is to be connected to the input/output (I/O) ports of the microcontroller board. Once they are properly connected, the circuit will then send an output signal upon detecting a moving living object.

In general, the PIR sensor can operate in two modes – the single triggering mode and repeat triggering mode. The jumper setting of the PIR sensor dictates which mode it is to operate. When the jumper is set to 'low', the sensor operates in the single triggering mode. On the other hand, when the jumper is 'high', the sensor works in the repeat triggering mode.

At the single triggering mode, the sensor output becomes 'high' (i.e. a logic 1 is sent out) for a certain duration upon detecting a moving object (which is usually a living being). It then changes to 'low' (i.e. a logic 0 or 0 V is sent out) and stays at that level for a fixed duration. After that, the output switches back to 'high' again if the object is still detected at the sensing area. Hence, the PIR sensor generates high/low pulses if an object is detected in continuous motion.

When the sensor is changed to the repeat triggering mode, it similarly sends out a 'high' output signal upon detecting a moving object. Unlike the single triggering mode, however, the sensor will always remain 'high' as long as the moving object is still detected. The output will only go 'low' when the object stops moving or leaves the sensing area. Hence, the PIR sensor only generates 'high' pulses when the object is in continuous motion.

2.3.2 Vision sensor module

The Pi camera (refer to Figure 2.5) is used in this project to detect and recognize the face of a moving object. In this project, the Pi camera is plugged directly into the camera serial interface (CSI) connector on the embedded system developer board. This camera module is chosen due to its tiny size, light weight (~3 g), and high sensor resolution (8 MP), thus making it easily to be installed onto the frame structure of robot head, as can be seen in Figure 2.6.

2.3.3 Servo motor

In general, a servo motor is a closed-loop transducer which converts electrical energy into mechanical energy – the output shaft rotates once it receives an input signal. It employs a closed-loop mechanism in which a positional feedback is used to control rotational speed and position. Since the robot head requires high torque to orientate its position, the TowerPro MG946R Metal Gear servo motor (refer to Figure 2.7) which

Figure 2.5 Raspberry Pi Camera V2.1.

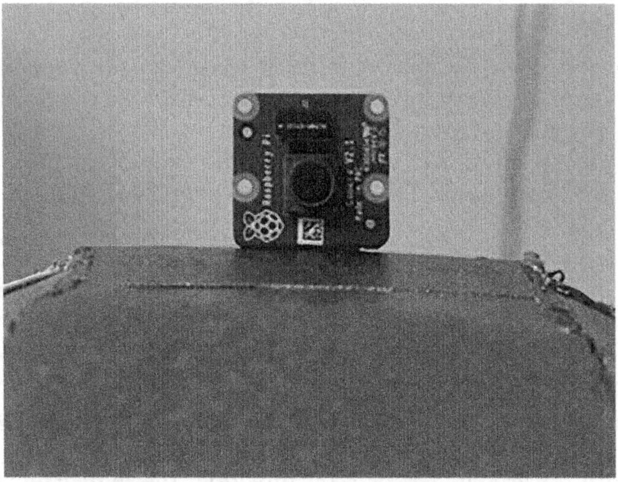

Figure 2.6 Installation of the Pi Camera onto the top of the robot head.

could exhibit a high holding torque of 13 kg/cm is selected for this project. As can be observed from Figure 2.7, the MG946R servo motor comes with connections made up of three wires – each of which is distinguished using different colors. The red wire is to be connected to a +5 V voltage source, the brown (or sometimes black) wire is to be connected to ground, and the orange (or sometime white) wire is used to receive pulse width modulation (PWM). Very often, the orange wire is connected directly to the output pin

Figure 2.7 The TowerPro MG946R servo motor.

of a microcontroller. The microcontroller feeds trains of voltage pulses to the servo motor via the orange wire, so as to control the rotational angle of the shaft. The angle of rotation of the servo motor depends on PWM. This is to say that the duration of a pulse fed to the input of the motor controls the angle the shaft turns. The frequency of the PWM signal is typically fixed.

In this project, the frequency of the MG946R servo motor was fixed at 50 Hz. At this frequency, the servo motor has a minimum and maximum pulse width of 0.5 and 2.5 ms, respectively. The position of the shaft when the pulse width varies is graphically illustrated in Figure 2.8. It is worthwhile noting that the position is not solely restricted to those shown in the figure. By carefully varying the pulse width, the shaft can turn to any specific angle which corresponds to the width. In order to allow the android's head to turn towards its left or right direction, the shaft is positioned at 90° when the head is at its neutral position. In other words, a 1.5 ms pulse width is fed to the servo motor every time the android's head is configured to face the front. The duty cycle of the shaft position is calculated by dividing the pulse width with the period of the PWM signal (which is 20 ms in this case). For example, the servo motor will set to 0° for a duty cycle of 2.5% (which is the percentage obtained when dividing 0.5 ms with 20 ms). Similarly, we shall get a duty cycle of 5% for 45°, 7.5% for 90°, 10% for 135°, and 12.5% for 180° rotation.

The servo motor is mounted to the base of the frame structure, so as to support the orientation of the android's head in either clockwise or counter

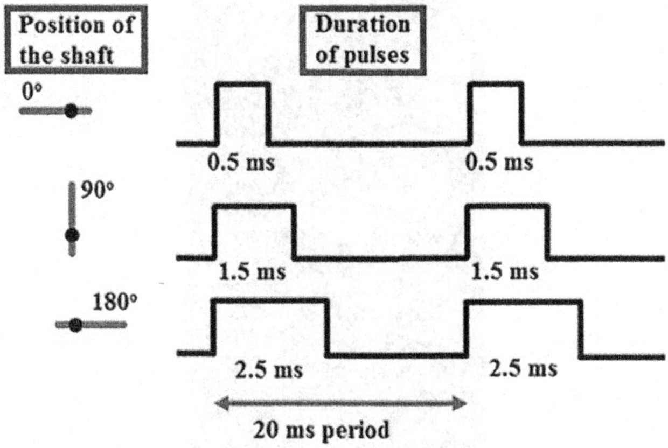

Figure 2.8 The rotational angle of the servo motor shaft when the pulse duration varies.

clockwise direction. When there is a person approaching the head from the right side, the head will rotate in the clockwise direction. Likewise, if someone approaches from the left, it will rotate in counter clockwise direction. When the person is standing right in front of the robot, the robot will return to its neutral position (i.e. a 1.5 ms pulse dictates the shaft to rotate 90°).

2.3.4 Raspberry Pi microcontroller

The Raspberry pi microcontroller is regarded as the 'brain' of the android's head. It is used to synchronize the activities of the android's head – (i) it coordinates the detection of the PIR sensors and the orientation of the servo motor and also (ii) performs facial detection and recognition upon receiving video streams from the vision sensor. Figure 2.9 tabulates the general purpose I/O (GPIO) pins configuration of the microcontroller, whereas Figure 2.10 depicts how the servo motor and PIR and vision sensors are interfaced to the microcontroller via a breadboard.

In order to interface the electronic circuits with the Raspberry Pi microcontroller, algorithms for matching the required specifications of this project are to be developed. In this case, a Python-based IDE (Integrated Development Environment), i.e. the Thonny editor which works on the Raspbian operating system (OS) was used to develop and test the algorithms. The Thonny editor is chosen because the tool consists of a built-in debugger which comes in handy when nasty bugs in a program are to be identified. Besides, the editor also allows sequential step-by-step coding evaluation, making it convenient for the user to check the syntax at every line. The interface of the Thonny editor in the Raspbian OS is shown in Figure 2.11.

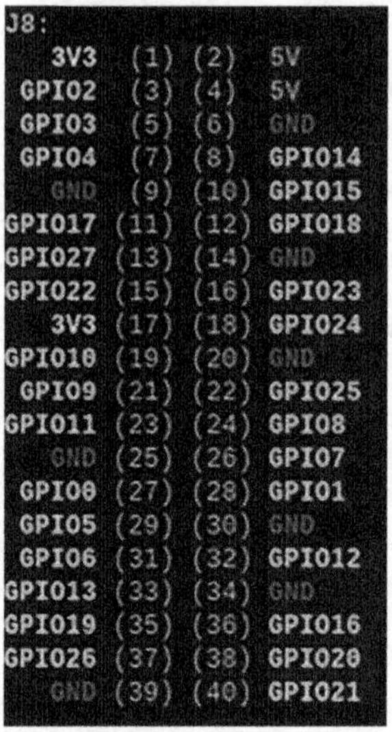

```
J8:
   3V3  (1) (2)  5V
 GPIO2  (3) (4)  5V
 GPIO3  (5) (6)  GND
 GPIO4  (7) (8)  GPIO14
   GND  (9) (10) GPIO15
GPIO17 (11) (12) GPIO18
GPIO27 (13) (14) GND
GPIO22 (15) (16) GPIO23
   3V3 (17) (18) GPIO24
GPIO10 (19) (20) GND
 GPIO9 (21) (22) GPIO25
GPIO11 (23) (24) GPIO8
   GND (25) (26) GPIO7
 GPIO0 (27) (28) GPIO1
 GPIO5 (29) (30) GND
 GPIO6 (31) (32) GPIO12
GPIO13 (33) (34) GND
GPIO19 (35) (36) GPIO16
GPIO26 (37) (38) GPIO20
   GND (39) (40) GPIO21
```

Figure 2.9 GPIO pins configuration of the Raspberry Pi.

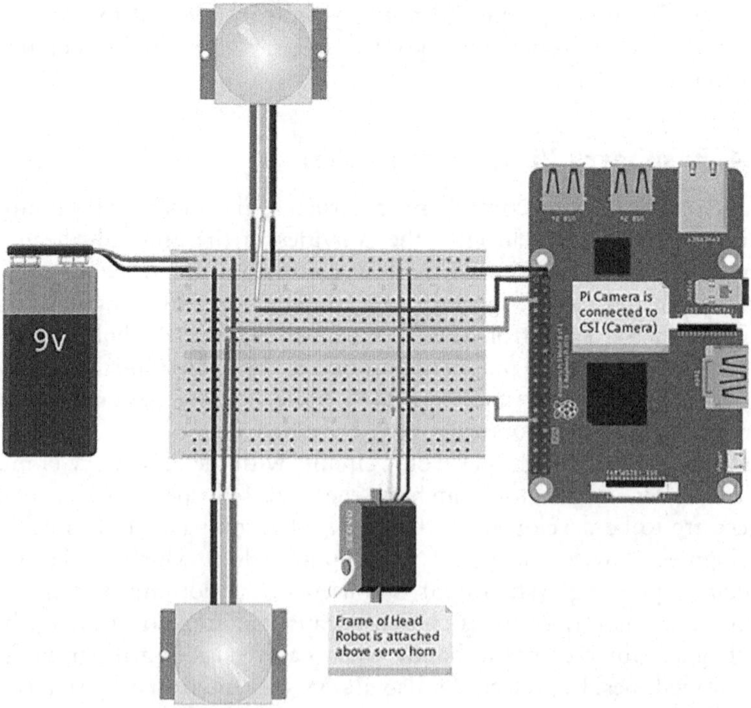

Figure 2.10 Breadboard layout diagram of Raspberry Pi microcontroller interfacing PIR and vision sensors and RC servo motor.

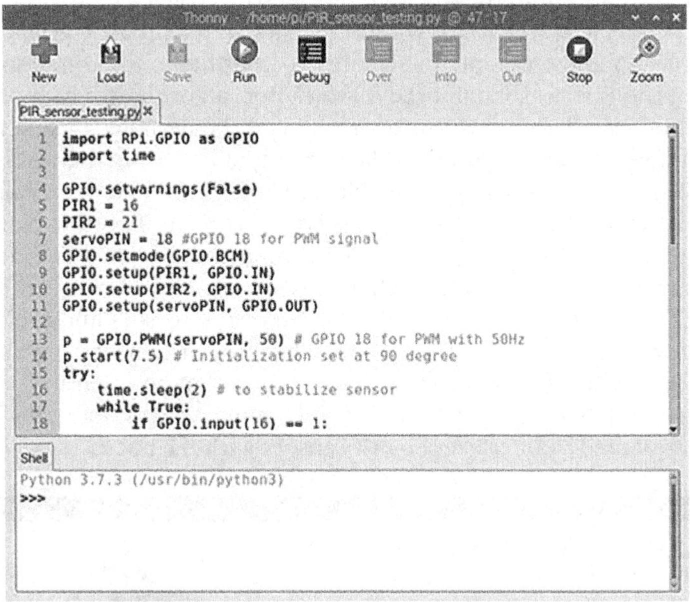

Figure 2.11 The Thonny editor.

2.4 PROGRAMMING THE ANDROID'S HEAD

The operation of the android's head can be classified into two phases – (i) it first tracks a moving object within its sensing distance (i.e. motion tracking) and then (ii) scans through the object's face to verify his/her identity (i.e. facial detection and recognition). These two phases are elaborated in detail in the sections below.

2.4.1 Motion tracking

The two PIR sensors attached to both sides of the robot's cheek allow it to determine whether the object is approaching from the left or right. Once the direction of the object is identified, the sensor which detects the object would then send a signal to the Raspberry Pi microcontroller. PWM signals are subsequently transmitted from the microcontroller to the servo motor to dictate its turning direction and its exact rotational angle.

The outputs of the sensors at the right and left sides are connected, respectively, to GPIO16 and GPIO21 of the microcontroller, whereas the orange wire of the servo motor is connected to any one of the four GPIO output pins: GPIO12 (pin 32), GPIO13 (pin 33), GPIO18 (pin 12), or GPIO19 (pin 35). The output pin is used to transmit the PWM signals to the servo motor. When the right sensor is triggered, the robot will rotate

to the right side. Similarly, when the left sensor is triggered, it will rotate to the left. Otherwise, the robot will rotate back to its initial center position.

In order to allow the microcontroller to coordinate the communication between the PIR sensors and the servo motor, algorithms which program the microcontroller to perform such task is to be developed using Python programming language. Some of the useful syntax are highlighted here.

To feed the output signals from the sensors to the microcontroller, the *GPIO.setup*(*X*, *GPIO.IN*) function is employed. This function configures GPIO16 and GPIO21 as input pins. The argument *X* in the function refers to the variables assigned to the right or left sensor. The delay function *time.sleep*(*Y*) is used to avoid multiple detection of the sensors and rapid servo rotation within a short period of time, where the argument *Y* refers to the delay time. Here, we have arbitrarily set $Y=5$ so that both sensors will only be activated at an interval of 5 s. Figure 2.12 summarizes some of the important syntax used to interface the PIR sensors with the microcontroller.

Coding	Diagram

```
p = GPIO.PWM(servoPIN, 50)
p.start(7.5) # Initial position before detection
try:
    time.sleep(2) # to stabilize sensor
```

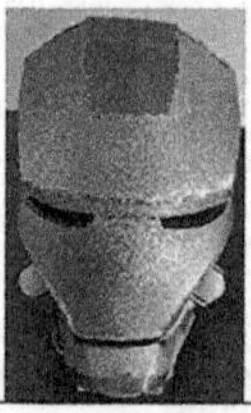

```
while True:
    if GPIO.input(16) == 1: # sensor1 detected
        p.ChangeDutyCycle(7.5)
        time.sleep(0.5)
        p.ChangeDutyCycle(5)
        time.sleep(0.5))
        p.ChangeDutyCycle(0)
        time.sleep(5)
```

```
    elif GPIO.input(21) == 1: # sensor2 detected
        p.ChangeDutyCycle(7.5)
        time.sleep(0.5)
        p.ChangeDutyCycle(10)
        time.sleep(0.5)
        p.ChangeDutyCycle(0)
        time.sleep(5)
```

```
    else: # no sensors detected
        p.ChangeDutyCycle(7.5)
        time.sleep(0.5)
        p.ChangeDutyCycle(0)
        time.sleep(3) #loop delay-less than detection delay
```

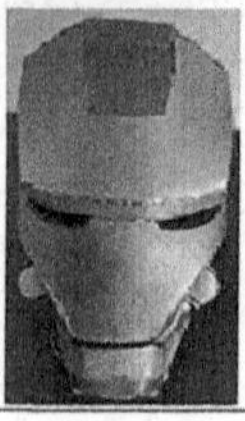

Figure 2.12 The syntax for PIR sensors detection.

When programming the microcontroller to transmit PWM pulses to the servo motor, the Python libraries listed below are employed:

i. The *GPIO.PWM(X, Y)* function is used to initiate the PWM signals, where *X* refers to the specific GPIO pin in which the control signal of the servo motor is connected to and *Y* refers to the frequency of the PWM signals (which is 50 Hz in this case).
ii. The *p.start(X)* function is used to initiate the position of the servo motor, where *X* refers to the duty cycle of the shaft position. Since the servo motor is initially configured to be at its center position, *X* is thus set to 7.5.
iii. The *p.ChangeDutyCycle(X)* function is used to send the specific pulse width to the servo motor so that the shaft could rotate to the desired position. Here, *X* refers to the value of the duty cycle which ranges from 2.5% to 12.5% (which corresponds to the 0° to 180° rotational angle).

The following three conditions are to be considered when designing the rotational angle of the servomotor:

i. When the head turns from its initial center to its right position, the servo motor is to rotate clockwise, with a pulse width of 0.5 ms fed to the motor.
ii. When the head turns from its initial center to its left position, the servo motor is to rotate counter clockwise, with a pulse width of 2.5 ms fed to the motor.
iii. When the head returns to its center position, a pulse width of 1.5 ms is to be fed to the servo motor.

Figure 2.13 summarizes the syntax used to control the movement of the servo motor for the three conditions mentioned above.

2.4.2 Facial detection and recognition

To equip the android's head with the capability to detect and recognize the countenance of its target, AI is to be programed into the microcontroller. In this case, machine learning has been implemented to develop the algorithm. In a nutshell, machine learning is nothing more than a set of learning algorithms which train the microcontroller to generate new skills based on inferences from a pool of existing training data, known as dataset. The types of machine learning can be broadly classified into supervised learning, unsupervised learning, semi-supervised learning, reinforcement learning, multitask learning, ensemble learning, neural network, and instance-based learning (Dey 2016). For the android's head project illustrated here, supervised learning which makes use of dataset to train the robot is employed.

<table>
<tr><th>Syntax</th><th>Illustration of the Head Movement</th></tr>
<tr><td>p = GPIO.PWM(servoPIN, 50)
p.start(7.5) #middle at the beginning
while True:
        p.ChangeDutyCycle(7.5)
        time.sleep(0.5)
        p.ChangeDutyCycle(0)</td><td>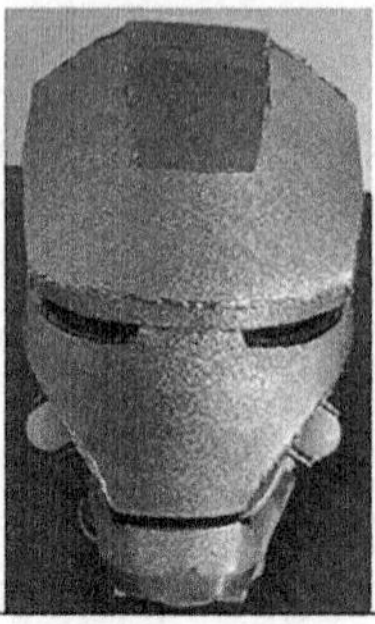</td></tr>
<tr><td>while True: #turns right
        p.ChangeDutyCycle(7.5)
        time.sleep(0.5)
        p.ChangeDutyCycle(5)
        time.sleep(0.5)
        p.ChangeDutyCycle(2.5)
        time.sleep(0.5)
        p.ChangeDutyCycle(0)</td><td></td></tr>
<tr><td>while True: #turns left
        p.ChangeDutyCycle(7.5)
        time.sleep(0.5)
        p.ChangeDutyCycle(10)
        time.sleep(0.5)
        p.ChangeDutyCycle(12.5)
        time.sleep(0.5)
        p.ChangeDutyCycle(0)</td><td>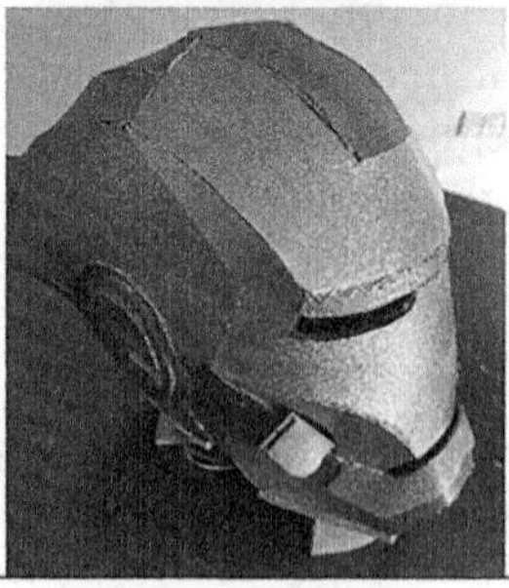</td></tr>
</table>

Figure 2.13 The key syntax which dictates the orientation of the android's head.

Before the target faces can be detected and recognized properly in video streams, the 128-dimension face embedding for each of different persons have to be computed using deep neural network. To quantify the face data into a dataset, a script is to be written with the required Python libraries such as *imutils*, *face_recognition*, and *OpenCV*.

To ease the process of debugging, command line arguments can be used at the beginning of the script. Command line arguments are flags given to a program or script so that different inputs can be assigned to a program without changing the original code. First, the argument parser is instantiated using the *argparse.ArgumentParser*() function. The argument parsers

are then constructed using the *add_argument*("-*X*", "--*Y*", *required* = *True*, *help* = "*additional information in terminal*") function, where *X* refers to the shorthand of the flag (only one character could be used) and *Y* refers to the longhand of the flag. To encode the face data from the dataset, three flags are to be constructed using the argument parser, namely *dataset* (i.e. path to dataset), *encodings* (i.e. the file that contains 128-d face encodings), and *detection-method* (i.e. the method for face detection, which could be either 'hog' or 'cnn'). In this project, the 'hog' method (Histogram of Oriented Gradients) was applied since it is less intensive in memory consumption. After the argument parsers are constructed, all the command line arguments are parsed using the *vars*(*.parse_args*()) function.

After all the essential arguments are defined, the paths in which the dataset is stored are linked using the *list*(*paths.list_images*(*args*["*dataset*"])) function. The list of known encodings and names are initialized using the *knownEncodings* and *knownNames* variables. A loop to process each of the different face data in the dataset is then created. The *enumerate*() function is used in this loop to accept the arguments and to run the loop iteratively. To extract the names of different individuals from the image path, the.*split*(*os.path.sep*) function is used. The *cv2.imread*() function is used to load the input image from the dataset, and the extracted image is then converted from BGR (OpenCV ordering) to RGB (dlib ordering) for further processing using the *cv2.cvtColor*(*X*, *cv2.COLOR_BGR2RGB*) function, where X refers to the input image.

To localize the faces in the input image, the *face_recognition.face_locations*(*X*, *model* = *args*["*detection_method*"]) function is used to trace the *x* and *y* coordinates of the bounding boxes around the detected face. Next, the *face_recognition.face_encodings*(*X*, *Y*) function is used to compute face embedding in the form of 128-d vectors based on the detection method applied, where X denotes the converted image and Y denotes the coordinates of the localized face in the image. The face embedding computed are then added to the list of known encodings and names. The *data* dictionary is initialized and constructed with the *encodings* and *names* keys using the *data* = {"*encodings*": *knownEncodings*, "*names*": *knownNames*} function. The data of facial encodings is then stored using the *f.write*() library so that they can be used in the facial recognition script later.

The command line below is to be typed at the command terminal to execute the process of face embedding on the Raspberry Pi microcontroller:

python encode_faces.py --dataset dataset --encodings encodings.pickle \ --detection-method hog

As can be seen in Figure 2.14, the *encodings.pickle* file is generated after the execution of the command line above. This file consists of the 128-dimension facial encodings for each face in the dataset. Figure 2.15 depicts a snapshot of the terminal when the process of face embedding is completed.

To recognize the faces in video streams, a main execution script is to be written with the required Python libraries such as imutils, face_recognition,

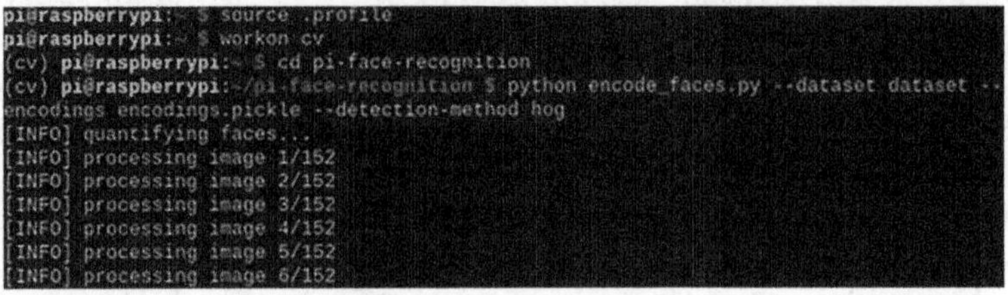

Figure 2.14 Command to execute face embedding.

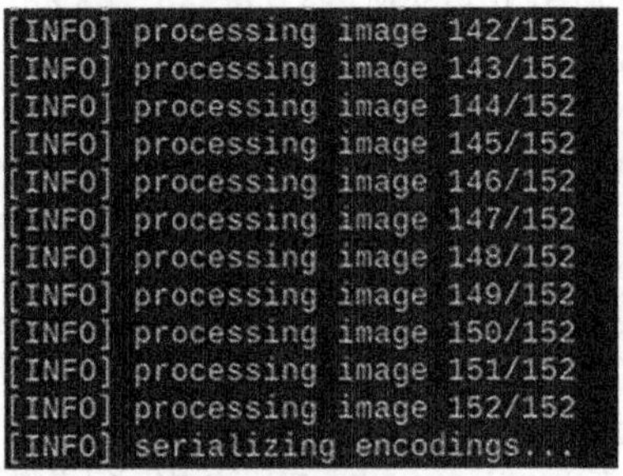

Figure 2.15 Completion of executing face embedding from the dataset.

and OpenCV. In this program, the Haar cascade model in OpenCV is used to detect and localize the face data in video streams. The Haar cascade model is an algorithm of object (or face) detection using machine learning with the ability to identify objects in a still image or video frame. This machine learning-based model works by training a cascade function from huge sets of positive and negative images. It is then applied to detect and identify required objects (or faces) in other images (Viola and Jones 2001).

To perform facial detection and recognition in video streams, two flags constructed using the argument parser, i.e. *cascade* (the path to the Haar cascade model in OpenCV) and *encodings* (the path to the serialized database containing 128-d facial encodings), are required. Before looping the frames in the video streams, the following objects are to be instantiated:

i. The *pickle.loads()* function is used to load the known faces and embeddings from *encodings.pickle.*
ii. The *cv2.CascadeClassifier()* function is used to instantiate the Haar cascade model for detecting faces.
iii. To initialize the video streams using the Pi camera, the *Videostream(usePiCamera = True)*.start() function is used. The FPS (frames per second) counter is initialized using the *FPS().start()* function.

Next, a loop to capture frames from the video streams is created. The captured frames are to be processed for facial recognition. In other words, they are to be matched with those in the dataset. Prior to performing facial recognition, however, the captured images have to be resized and the color format has to be converted first. The frames are captured from the camera using the *vs.read*() function, and they are resized to 500 pixels so as to speed up processing time. The *imutils.resize*(*X*, *width* = *Y*) function is employed for resizing. The *X* and *Y* arguments in the function denote, respectively, the captured frame and the size of the new frame which corresponds to its resolution, i.e. *Y* is expressed in pixels. To perform facial detection on the captured frames, the images are converted from BGR to grayscale using the *cv2.COLOR_BGR2GRAY* function. On the other hand, for facial recognition, the images are converted from BGR to RGB using the *cv2.COLOR_BGR2RGB* function.

To detect the faces in the grayscale format, the *rects=detector.detectMultiScale*(X, *scaleFactor* = 1.2, *minNeighbors* = 5, *minSize* = (30, 30)) function is used. In the function, the argument X refers to the grayscale image, *scaleFactor* specifies the scale of reduction that the image has undergone, *minNeighbors* specifies how many surrounding neighbors the candidate rectangle should have to retain it, and *minSize* refers to the minimum possible object (or face) size that is to be detected (Bae et al. 2020). The rescaling of the captured frames (from larger to smaller size) is important to enable the faces in the image to be detectable. The smaller is the step of resizing, the higher is the accuracy of matching the captured faces with those in the facial embedding. The high accuracy, however, comes with the expense of a slower detection rate. On the contrary, a larger resizing step results in a faster detection rate, but it increases the risk of missing some fine details in the image and this may then diminish the accuracy of detection.

In facial recognition, the *minNeighbors* argument is of particular importance. This is because the number of neighbors detected near the candidate rectangle affects the quality of the detected faces. For optimized detection, the values assigned to *minNeighbors* usually fall within the range of 3–6. The results of the detected faces are stored in the *rects* dictionary. This is to say that the *rects* dictionary consists of a list of face bounding boxes which corresponds to the face locations in the image. The coordinates of the detected face bounding boxes are then converted and reordered using the *boxes* = [(*y*, *x* + *w*, *y* + *h*, *x*) *for* (*x*, *y*, *w*, *h*) *in rects*] function. This step is necessary since OpenCV gives the coordinates of the bounding box in the (*x*, *y*, *w*, *h*) format, while, the format of (top, right, bottom, left) is needed for facial detection. For each of the different face bounding boxes, the 128-d facial encodings are computed using the *face_recognition.face_encodings*(X, *Y*) function, where *X* refers to the captured frame in RGB format and *Y* refers to the face bounding box in reordered coordinates.

The subsequent step is to call a loop for identifying the faces and checking for matches from the facial encodings. The *face_recognition.*

compare_faces(*data*[“*encodings*”], *encoding*) function is used for this purpose. If matches are found, the *count* dictionary is initialized to calculate the frequency each face is matched to the known encodings. To recognize whose face it most likely is, a polling system is applied in this face recognition system. All the matched indexes are looped iteratively, and in every iteration, the number of counts would be incremented for each recognized face. To determine the recognized face with the highest count of polls, the *name* = *max*(*counts*, *key* = *counts.get*) function is used. In the event, an unlikely tie occurs (i.e. the same number of votes between two or more entries), and the first entry (i.e. the name of the first person) in the dictionary will be selected.

Once the detected face in the rectangular bounding box is recognized, the predicted name of the person is displayed via video streams. To draw rectangles surrounding each detected face, the *cv*2.*rectangle*(*frame*, (*left*, *top*), (*right*, *bottom*), (0, 255, 0), 2) function is used, In this function, the argument *frame* represents the captured frame from the video streams, the (*left*, *top*) argument indicates the starting point of the rectangular box, the (*right*, *bottom*) argument indicates the ending point of rectangular box, the (0, 255, 0) color code coordinate indicates that the border of the rectangular box is in green, and the value 2 at the last part of the function represents the thickness of the border in pixels. To display the predicted name of the person, the *cv*2.*putText*(*frame*, *name*, (*left*, *y*), *cv*2.*FONT_HERSHEY_SIMPLEX*, 0.75, (255, 0, 0), 2) function is used. Here, the (*left*, *y*) argument indicates the position of the name to be displayed, the *cv*2.*ONT_HERSHEY_SIMPLEX* refers to the font type, 0.75 refers to the font scale, the (255, 0, 0) color code coordinate indicates that the color of the text is red, and the value 2 at the last part of the function indicates the thickness of the font in pixels. As shown in Figure 2.16, the following command is executed in the terminal to initialize facial detection and recognition on the video streams:

python pi_face_recognition.py --cascade haarcascade_frontalface_default.xml \ --encodings encodings.pickle

When the *q* key is pressed on the keyboard, the video streams will be halted and the results of elapsed time and approximate FPS would be displayed in the terminal. Some examples of detected faces are illustrated in Figure 2.17.

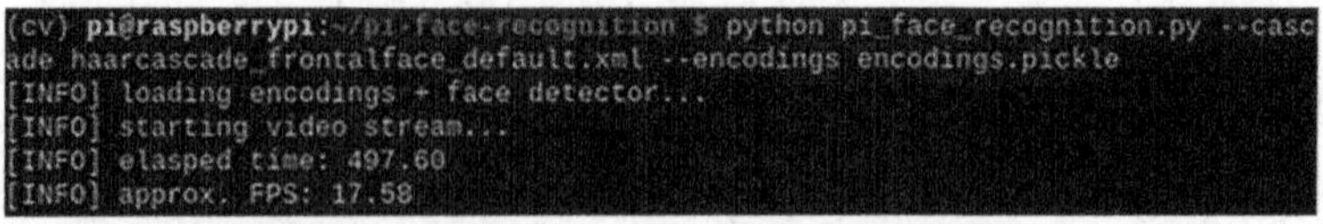

Figure 2.16 Command to execute facial detection and recognition.

Figure 2.17 The predicted names are displayed in the video streams when the robot recognizes the faces.

2.5 CHALLENGES

In this section, some of the impediments faced throughout the development of this project are highlighted. First, the Raspberry Pi microcontroller could be easily overheated when it was performing facial detection and recognition. Based on the observation, the temperature of the CPU could easily rise to 70–80°C when it was heavily loaded. The Raspberry Pi microcontroller could be throttled at the temperature of around 90°C when the microcontroller operated continuously for a long period of time. In order to cope with this issue, a thermal cooling system with dual fans and thermal paste is attached to the CPU of the Raspberry Pi microcontroller. Once the active cooling system, as shown in Figure 2.18, was installed, thermal heat was dissipated effectively, and the temperature of the CPU is controlled within the range of 50°C–60°C throughout its operation.

The second issue that we faced was the insufficient supply of power. The servo motor initially received its power directly from the Raspberry Pi microcontroller, and it worked perfectly fine when there was no load attached to it. Once the android's structure was mounted onto the servo motor, however, we realized that the motor failed to generate sufficient torque to turn its shaft. We, therefore, resorted to an external power source

Figure 2.18 Installation of an active cooling system onto the Raspberry Pi microcontroller.

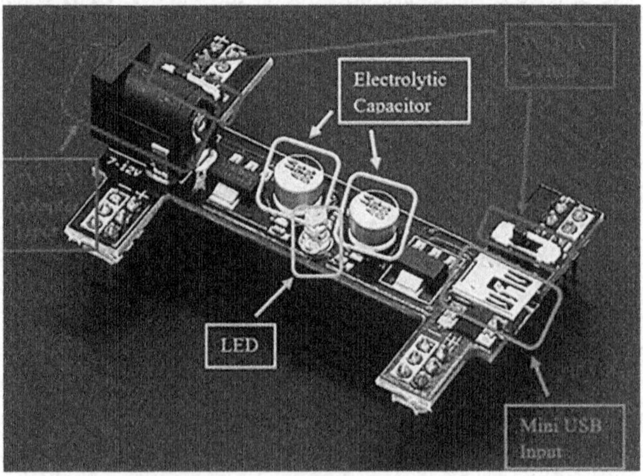

Figure 2.19 An external power supply module.

to fuel the servo motor. Figure 2.19 depicts the separate power module used for the servo motor. As can be observed from the figure, the module consists of two voltage inputs, viz. a 9 V battery and a 12 V/2 A power adapter. This power supply module is able to provide a 5 V and a 3.3 V linear regulated output voltage. Two slide switches at both edges of the module allow the user to select the appropriate output voltage level which fits

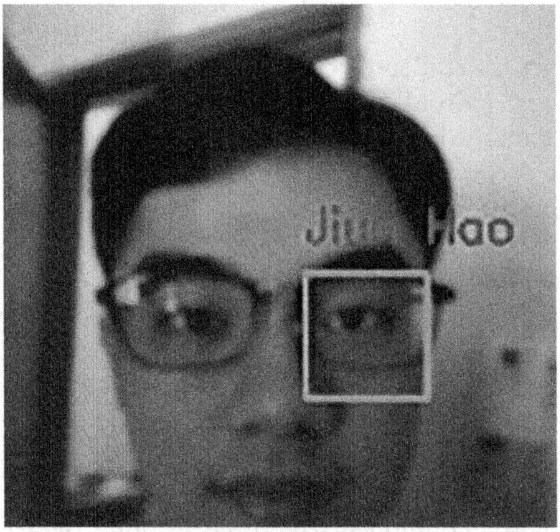

Figure 2.20 A fraction of the face would be detected as an individual object when *scaleFactor* is set too low.

the application. The electrolytic capacitors on the board are used as ripple filtering and to provide stable voltage over a long period of usage (Cytron Technologies Malaysia 2019).

The last and also the most strenuous issue faced in the development phase of the robot was programming the machine learning algorithms into the microcontroller. When executing the.*detectMultiScale*() function for facial detection, extra care has to be devoted to the *scaleFactor* and *minNeighbors* arguments as they affect significantly the speed and accuracy of the facial detection process. The *scaleFactor* was initially set to 1.05 with the intention of improving the quality of the detected faces. The problem which ensued from this low parameter was that the eyes or skin (i.e. a small portion of the face) could sometimes be detected as an individual object, as shown in Figure 2.20. After a series of careful parametric adjustment, the *scaleFactor* was set to 1.2 which eventually gives a sufficiently accurate detection of the faces. Similarly, *minNeighbors* was initially set to 10 since a high value was expected to give highly accurate facial detection result. It was found, however, that it took too long for the machine learning algorithm to complete its facial detecting process. To improve the speed of the process and, at the same time, to retain the quality of the detected faces, we decided to resort to half of the initial value (i.e. *minNeighbors* was assigned 5). In order to keep the average time for facial detection and recognition to be within 0.2 s, the pace of facial recognition is to be accelerated as well. To do so, the captured images were resized to a resolution of 500 pixels. The parameter keeps the resolution reasonably low, without sacrificing the accuracy of face recognition.

2.6 CONCLUSIONS

This chapter gives a detailed elucidation on the design, development, and construction of an android's head. Since an android is a humanoid robot which could imitate the behavior of a real human being, the android's head is equipped with certain intelligence. In a nutshell, the robot proposed here is able to perform the following functions:

i. The robot is able to sense and trace a moving object. Once it has identified the object, it will then orientate its position to face the moving target.
ii. Like human beings, 'memory' is programmed into the robot. Hence, it could recollect from its memory for familiar faces so as to check if the target is someone that it recognizes.

Some of the essential components required to build the robot are a servo motor, a pair of PIR sensors, a vision sensor, and a microcontroller.

APPENDIX

A. ALGORITHM FOR FACIAL DETECTION

```
#import the necessary packages
  import RPi.GPIO as GPIO
  import simpleaudio as sa
  import time

  #setup the GPIO pins for input and output
  GPIO.setwarnings(False)
  PIR1=16 #GPIO16 for right PIR sensor
  PIR2=21 #GPIO21 for left PIR sensor
  servoPIN=18 #GPIO 18 for PWM signal
  GPIO.setmode(GPIO.BCM)
  GPIO.setup(PIR1, GPIO.IN) #configure GPIO16 as input
  GPIO.setup(PIR2, GPIO.IN) #configure GPIO21 as input
  GPIO.setup(servoPIN, GPIO.OUT) #configure GPIO18 as output
  p=GPIO.PWM(servoPIN, 50) #GPIO 18 for PWM with 50Hz
  p.start(7.5) #Initialization set at 90 degree

  #Play the welcoming message of the robot
  wave_obj1=sa.WaveObject.from_wave_file("/home/pi/Desktop/Voice
Output/Robot_Intro.wav")
  play_obj1=wave_obj1.play()
  play_obj1.wait_done()
```

```
time.sleep(2)

try:
    time.sleep(2) #Delay to stabilize sensor

    #Loop for human detection (motion)
    while True:

        #For detection by right PIR sensor
        if GPIO.input(16) == 1:
            p.ChangeDutyCycle(7.5) #90 degree
            time.sleep(0.5)
            p.ChangeDutyCycle(5) #67.5 degree
            time.sleep(0.5)
            p.ChangeDutyCycle(2.5) #45 degree
            time.sleep(0.5)
            p.ChangeDutyCycle(0) #Stop sending inputs to servo
            print("Motion 1 Detected...")
             wave_obj2=sa.WaveObject.from_wave_file("/home/pi/Desktop/
Voice Output/Robot_Right.wav")
              play_obj2=wave_obj2.play() #Output sound when robot turns
right
            play_obj2.wait_done()
            time.sleep(5) #Delay to avoid multiple detection

        #For detection by left PIR sensor
        elif GPIO.input(21) == 1:
            p.ChangeDutyCycle(7.5) #90 degree
            time.sleep(0.5)
            p.ChangeDutyCycle(10) #112.5 degree
            time.sleep(0.5)
            p.ChangeDutyCycle(12.5) #135 degree
            time.sleep(0.5)
            p.ChangeDutyCycle(0)
            print("Motion 2 Detected...")
            wave_obj3=sa.WaveObject.from_wave_file("/home/pi/Desktop/
Voice Output/Robot_Left.wav")
          play_obj3=wave_obj3.play() #Output sound when robot turns left
          play_obj3.wait_done()
          time.sleep(5)

        #No detection by both sensors (back to center)
        else:
            p.ChangeDutyCycle(7.5) #90 degree
            time.sleep(0.5)
```

```
            p.ChangeDutyCycle(0)
            print("No Motion Detected")
            wave_obj4=sa.WaveObject.from_wave_file("/home/pi/Desktop/
Voice Output/Robot_Front.wav")
             play_obj4=wave_obj4.play() #Output sound when robot turns
back to front (center)
            play_obj4.wait_done()
            time.sleep(3) #Loop delay (less than detection delay)

   #Stop the operation in the event of any input from keyboard
   except KeyboardInterrupt:
   p.stop()
   GPIO.cleanup()
```

B. ALGORITHM TO CONVERT EACH FACE TO FACE EMBEDDING

```
# import the necessary packages
from imutils import paths
import face_recognition
import argparse
import pickle
import cv2
import os

# construct the argument parser and parse the arguments
ap=argparse.ArgumentParser()
ap.add_argument("-i", "--dataset", required=True,
          help="path to input directory of faces+images")
ap.add_argument("-e", "--encodings", required=True,
          help="path to serialized db of facial encodings")
ap.add_argument("-d", "--detection-method", type=str, default="cnn",
help="face detection model to use: either `hog` or `cnn`")
args=vars(ap.parse_args())

# grab the paths to the input images in our dataset
print("[INFO] quantifying faces...")
imagePaths=list(paths.list_images(args["dataset"]))

# initialize the list of known encodings and known names
knownEncodings=[]
knownNames=[]
```

```
# loop over the image paths
for (i, imagePath) in enumerate(imagePaths):
        # extract the person name from the image path
        print("[INFO] processing image {}/{}".format(i+1,
                len(imagePaths)))
        name=imagePath.split(os.path.sep)[-2]

        # load the input image and convert it from RGB (OpenCV
ordering)
        # to dlib ordering (RGB)
        image=cv2.imread(imagePath)
        rgb=cv2.cvtColor(image, cv2.COLOR_BGR2RGB)

        # detect the (x, y)-coordinates of the bounding boxes
        # corresponding to each face in the input image
        boxes=face_recognition.face_locations(rgb,
                model=args["detection_method"])

        # compute the facial embedding for the face
        encodings=face_recognition.face_encodings(rgb, boxes)

        # loop over the encodings
        for encoding in encodings:
            # add each encoding+name to our set of known names and
            # encodings
            knownEncodings.append(encoding)
             knownNames.append(name)

# dump the facial encodings+names to disk
print("[INFO] serializing encodings...")
data={"encodings": knownEncodings, "names": knownNames}
f=open(args["encodings"], "wb")
f.write(pickle.dumps(data))
f.close()
```

C. ALGORITHM FOR FACIAL RECOGNITION

```
# import the necessary packages
from imutils.video import VideoStream
from imutils.video import FPS
import face_recognition
import argparse
import imutils
import pickle
```

```
import time
import cv2

# construct the argument parser and parse the arguments
ap=argparse.ArgumentParser()
ap.add_argument("-c", "--cascade", required=True,
        help="path to where the face cascade resides")
ap.add_argument("-e", "--encodings", required=True,
        help="path to serialized db of facial encodings")
args=vars(ap.parse_args())

# load the known faces and embeddings along with OpenCV's Haar
# cascade for face detection
print("[INFO] loading encodings+face detector...")
data=pickle.loads(open(args["encodings"], "rb").read())
detector=cv2.CascadeClassifier(args["cascade"])

# initialize the video stream and allow the camera sensor to warm up
print("[INFO] starting video stream...")
vs=VideoStream(src=0).start()
# vs=VideoStream(usePiCamera=True).start()
time.sleep(2.0)

# start the FPS counter
fps=FPS().start()

# loop over frames from the video file stream
while True:
        # grab the frame from the threaded video stream and resize it
        # to 500px (to speedup processing)
        frame=vs.read()
        frame=imutils.resize(frame, width=500)

        # convert the input frame from (1) BGR to grayscale (for face
        # detection) and (2) from BGR to RGB (for face recognition)
        gray=cv2.cvtColor(frame, cv2.COLOR_BGR2GRAY)
        rgb=cv2.cvtColor(frame, cv2.COLOR_BGR2RGB)

        # detect faces in the grayscale frame
        rects=detector.detectMultiScale(gray, scaleFactor=1.1,
                minNeighbors=5, minSize=(30, 30),
                flags=cv2.CASCADE_SCALE_IMAGE)

        # OpenCV returns bounding box coordinates in (x, y, w, h) order
        # but we need them in (top, right, bottom, left) order, so we
        # need to do a bit of reordering
```

```
boxes=[(y, x+w, y+h, x) for (x, y, w, h) in rects]

# compute the facial embeddings for each face bounding box
encodings=face_recognition.face_encodings(rgb, boxes)
names=[]

# loop over the facial embeddings
for encoding in encodings:
    # attempt to match each face in the input image to our known
    # encodings
    matches=face_recognition.compare_faces(data["encodings"],
            encoding)
    name= "Unknown"

    # check to see if we have found a match
    if True in matches:
            # find the indexes of all matched faces then initialize a
            # dictionary to count the total number of times each face
            # was matched
            matchedIdxs=[i for (i, b) in enumerate(matches) if b]
            counts={}

            # loop over the matched indexes and maintain a count for
            # each recognized face face
            for i in matchedIdxs:
                    name=data["names"][i]
                    counts[name]=counts.get(name, 0)+1

            # determine the recognized face with the largest number
            # of votes (note: in the event of an unlikely tie Python
            # will select first entry in the dictionary)
            name=max(counts, key=counts.get)

    # update the list of names
    names.append(name)

# loop over the recognized faces
for ((top, right, bottom, left), name) in zip(boxes, names):
        # draw the predicted face name on the image
        cv2.rectangle(frame, (left, top), (right, bottom),
                (0, 255, 0), 2)
        y=top - 15 if top - 15>15 else top+15
        cv2.putText(frame, name, (left, y), cv2.
FONT_HERSHEY_SIMPLEX,
                0.75, (0, 255, 0), 2)
```

```
        # display the image to our screen
        cv2.imshow("Frame", frame)
        key=cv2.waitKey(1) & 0xFF

        # if the `q` key was pressed, break from the loop
        if key == ord("q"):
                break

        # update the FPS counter
        fps.update()

# stop the timer and display FPS information
fps.stop()
print("[INFO] elasped time: {:.2f}".format(fps.elapsed()))
print("[INFO] approx. FPS: {:.2f}".format(fps.fps()))

# do a bit of cleanup
cv2.destroyAllWindows()
vs.stop()
```

ACKNOWLEDGMENT

This work was supported in part by the UTAR research fund (project: IPSR/RMC/UTARRF/2019-C1/Y01).

REFERENCES

Bae, J. P., S. Yoon, M. Vania, and D. Lee. 2020. Three dimensional microrobot tracking using learning-based system. *International Journal of Control, Automation and Systems* 18: 21–28.

Cytron Technologies Malaysia. 2019. Black wings-3.3v/5v power breadboard adapter BDP. [online] Available at: https://my.cytron.io/p-black-wings-3.3v-5v-power-breadboard-adapter-bdp?src=account.order (Accessed 25 February 2020).

Dey, A. 2016. Machine learning algorithms: A review. *International Journal of Computer Science and Information Technologies* 7: 1174–1179.

Ho, Y. K. 2004. Design and implementation of a mobile robot. BEng. Thesis, Universiti Teknologi Petronas, Malaysia.

IEEE. 2019. Robots. IEEE spectrum. Available at: robots.ieee.org/learn/ (accessed April 30, 2020).

JoshuaGuess. 2016. Build an iron man helmet for cheap! [online] Available at: https://www.instructables.com/id/Build-an-Iron-Man-Helmet-for-Cheap/ (Accessed 25 February 2020).

Viola, P. and M. Jones. 2001. Rapid object detection using a boosted cascade of simple features. *Proceedings of the Conference on Computer Vision and Pattern Recognition*, Kauai, USA.

REFERENCES FOR ADVANCE/FURTHER READING

Chong, K. H., S. P. Koh, S. K. Tiong, and K. H. Yeap. 2011. Design and development of automated digital circuit structure base on evolutionary algorithm method. *International Journal of Electronics, Computer, and Communications Technologies* 2: 1–8.

En, O. C., T. P. Chiong, Y. K. Ho, and L. S. Chyan. 2013. Flexible optical pulse generation for laser micromachining using FPGA board with integrated high speed DAC. *I-Manager's Journal on Digital Signal Processing* 1: 24–29.

Karmila, K., C. K. Hen, T. S. Kiong, and Y. K. Ho. 2011. Application of crossover factor on FPSGA in optimization problem. *Proceedings of the 2011 International Conference on Information and Intelligent Computing*, Hong Kong.

Lai, K. C., S. K. Lim, P. C. Teh, and K. H. Yeap. 2016. Modeling electrostatic separation process using artificial neural network (ANN). *Procedia Computer Science* 91: 372–381.

Lee, W. T., H. Nisar, A. S. Malik, and K. H. Yeap. 2013. A brain computer interface for smart home control. *Proceedings of the 17th International Symposium on Consumer Electronics*, Hsinchu, Taiwan.

Nisar, H., H. Z. Yang, and Y. K. Ho. 2015. Automated fruit and flower counting using digital image analysis. *Proceedings of the 4th International Conference on Computer Engineering and Mathematical Sciences*, Langkawi, Malaysia.

Nisar, H., H. Z. Yang, and Y. K. Ho. 2015. Predicting yield of fruit and flowers using digital image analysis. *Indian Journal of Science and Technology* 8: 1–6.

Nisar, H., H.-W. Khow, and K.-H. Yeap. 2018. Brain-computer interface: Controlling a robotic arm using facial expressions. *Turkish Journal of Electrical Engineering and Computer Sciences* 26: 707–720.

Nisar, H., M. B. Khan, W. T. Yi, Y. K. Ho and L. K. Chun. 2016. A non invasive heart rate measurement system for multiple people in the presence of motion and varying illumination. *International Journal of Disease Control and Containment for Sustainability* 1: 1–11.

Nisar, H., Z. Y. Lim, and K. H. Yeap. A simple non-invasive automated heart rate monitoring system using facial images. pp. 100–122. In: W. B. A. Karaa and N. Dey. [eds.] 2015. *Biomedical Image Analysis and Mining Techniques for Improved Health Outcomes*. IGI Global, Hershey, PA.

Ong, C. E., K. H. Yeap, S. C. Lee, and P. C. The. 2013. Flexible pulse-width tunable 1.06 μm pulsed laser source utilizing a FPGA based waveform generator. *Proceedings of the 4th IEEE International Conference on Photonics*, Melaka, Malaysia.

Rizman, Z. I., J. Adnan, F. R. Hashim, I. M. Yassin, A. Zabidi, F. K. Zaman, and K. H. Yeap. 2018. An improved controller for grass cutting application. *Journal of Fundamental and Applied Sciences* 10: 806–815.

Rizman, Z. I., J. Adnan, F. R. Hashim, I. M. Yassin, A. Zabidi, F. K. Zaman, and K. H. Yeap. 2018. Development of hybrid drone (HyDro) for surveillance application. *Journal of Fundamental and Applied Sciences* 10: 816–823.

Rizman, Z. I., F. R. Hashim, I. M. Yassin, A. Zabidi, F. K. Zaman, and K. H. Yeap. 2018. Design an electronic mouse trap for agriculture area. *Journal of Fundamental and Applied Sciences* 10: 824–831.

Rizman, Z. I., F. R. Hashim, I. M. Yassin, A. Zabidi, F. K. Zaman, and K. H. Yeap. 2018. Smart multi-application energy harvester using Arduino. *Journal of Fundamental and Applied Sciences* 10: 689–704.

Rizman, Z. I., M. T. Ishak, F. R. Hashim, I. M. Yassin, A. Zabidi, F. K. Zaman, K. H. Yeap, and M. N. Kamarudin. 2018. Design a simple solar tracker for hybrid power supply. *Journal of Fundamental and Applied Sciences* 10: 333–346.

Rizman, Z. I., M. T. Ishak, F. R. Hashim, I. M. Yassin, A. Zabidi, F. K. Zaman, K. H. Yeap, and M. N. Kamarudin. 2018. SPOSOL: 5 in 1 smart portable solar light. *Journal of Fundamental and Applied Sciences* 10: 347–364.

Rizman, Z. I., K. H. Yeap, N. Ismail, N. Mohamad, and N. H. R. Husin. 2013. Design an automatic temperature control system for smart electric fan using PIC. *International Journal of Science and Research* 2: 1–4.

Yeap, K. H., K. W. Thee, K. C. Lai, H. Nisar, and K. C. Krishnan. 2018. VLSI circuit optimization for 8051 MCU. *International Journal of Technology* 9: 142–149.

Chapter 3

Detecting DeepFakes for future robotic systems

Eric Tjon, Melody Moh, and Teng-Sheng Moh
San Jose State University

CONTENTS

3.1 INTRODUCTION

Deep learning-based generative models allow an alternative way to produce synthetic images and videos. Generative models have multiple applications such as creating a synthetic dataset or automatically performing image translations. These models are also used for art and entertainment. One application can create synthetic digital art from sketches. As these

generative models get better, the output becomes harder to distinguish from reality for both humans and machines alike.

Unfortunately, these generative models are also a cause for concerns when used for the wrong purpose. Algorithms commonly known as DeepFakes are one such threat. DeepFakes can generate and manipulate faces and audio in videos. Originally, DeepFakes were used for entertainment purposes such as editing movie scenes with different actors. The concern arises when these tools are used for deception and misinformation. Potentially, these tools can imitate political figures and trusted professionals. Since DeepFake algorithms are publicly available online, malicious users can easily acquire the means to carry out such deception.

Future robotic systems are vulnerable to DeepFake-based attacks as well. Video conferencing with trusted people such as doctors and lawyers can be imitated. Furthermore, any AI-based professional in the future can have their appearance wrongly synthesized through DeepFakes. These attacks can have long-lasting effects on the public's trust in digital media, including those used in the future robotic systems. Automated detection methods are therefore needed to combat these threats.

3.2 IMAGE SYNTHESIS AND DEEPFAKE TECHNOLOGIES

Video forgery involves altering a video from its original content. Different video manipulation methods exist to change the identity of a person in digital media. Earlier versions of these techniques require significant knowledge and effort to produce quality results. Recent advancements in deep learning tools are more automated and produce more realistic images. More specifically, conditional generative adversarial networks (GANs) have been shown to achieve realistic image synthesis (Mirza and Osindero 2014; Isola et al. 2017). These computer-generated images are difficult to distinguish from photographic images even for human observers. The ease and availability of deep-learning-based facial manipulation increases the concern that fake images and fake videos spread through social media and other mass media, which may quite easily deceive the general public.

3.2.1 Generative adversarial networks

GANs are generative models created with deep neural networks (Goodfellow et al. 2014). A generative model can take a sample of data and replicate its distribution either explicitly with an output of a function or implicitly with the output of synthetic data. GANs primarily target the latter function and generate samples that mimic the training data. Generative models are particularly suited for many tasks including creating synthetic datasets and manipulating images. GANs improve on many aspects of older

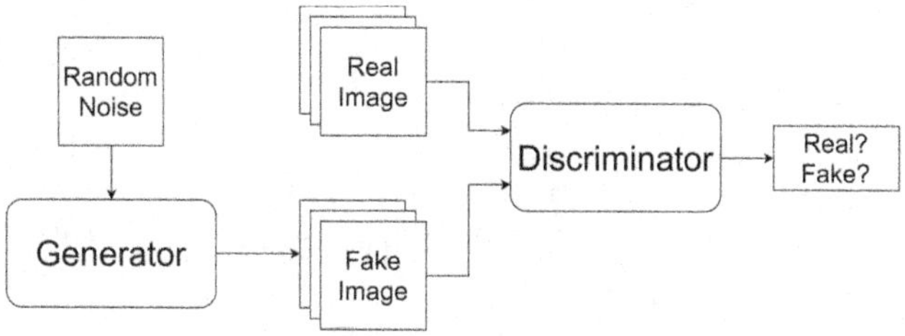

Figure 3.1 GAN architecture.

generative models. They are able to emulate complex data samples and generate output efficiently once trained. Furthermore, the output of a GAN is subjectively more realistic than other generative methods.

The base architecture of a GAN is comprised of a generator and discriminator (Figure 3.1). These two parts are implemented with deep neural networks, such as a multilayered perceptron. The generator is a network that is trained to generate samples of data. The input to the generator is a noise vector from a uniform distribution. The random noise allows the model to create varied output and capture the whole distribution of the original data. The discriminator is a classification model trained to distinguish between real and generated instance of data. The input is a single sample of data. The output is a classification score between 0 and 1 representing real or fake.

Adversarial training forms the basis of training a GAN (Goodfellow et al. 2014). The generator and discriminator view each other as adversaries; they try to minimize opposite loss functions, so they have opposing goals. In the training process, these two models take alternating steps to update and improve. In one step, the generator's weights are updated based on how well it can fool the discriminator. In the other step, the discriminator updates its weights based on how well it can identify generated data.

Conditional GANs are a special type of GAN that adds a condition to both the generator and discriminator (Mirza and Osindero 2014). This condition takes the form of a vector which goes into the input layer of the neural networks (Figure 3.2). The condition represents additional information about the data, such as a class label, category, or related image. Conditional GANs allow greater control over the output of the model. This feature is especially useful in image synthesis, where a specific image transformation is needed.

Conditions specify a specific category of data for the GAN to generate. For instance, a plain GAN trained on generating hand-written numbers would produce random numbers based on the random noise vector that serves as input. A conditional GAN could provide a class label to generate a specific number. Advancements in conditional GANs enable models to work with many types of conditions including text descriptions, semantic images, and videos.

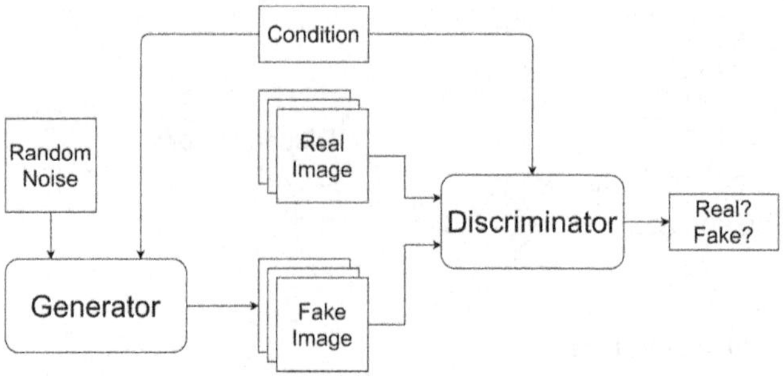

Figure 3.2 Conditional GAN architecture.

3.2.2 GAN image models

Deep learning techniques can also alter faces with a high-accuracy reconstruction, mainly relying on the technology of GANs. These techniques automatically learn how to generate faces in a certain pose, eliminating the need for a user to manually edit and blend in facial images.

3.2.2.1 General image synthesis

Conditional GANs can be designed to produce a variety of images. The following architectures are designed for creating images in a general category and are flexible with the type of data used.

Pix2pix is a conditional GAN architecture designed for image-to-image transformations (Isola et al. 2017). It takes in an image as the condition and results in a related, transformed image. This architecture is flexible and can learn many types of image functions such as black and white to color, semantic masks to real images, and edges to objects. In general, the condition image presented to the model is an encoding of a scene, and the output image is a realistic rendering of the same scene.

This model is trained in a supervised process, where it learns the transformation from input–output pairs in the training set. Therefore, this model can only learn transformations where existing examples can be obtained. With the training set's examples, the generator learns how to map a condition to the output through adversarial learning.

Pix2pix also introduced numerous improvements to the generator model and training loss (Isola et al. 2017). Since the condition image is an image with similar resolution and structure, the model can use this information to successfully generate the output image. This fact is reflected in the architecture of pix2pix. Typically, the generator of a GAN is a paired encoder and decoder. The input image is encoded into features through the encoder. These features are decoded into an output image by the decoder. To preserve overall structure and context from the input image, pix2pix implements skip connections that link similarly sized encoder and decoder layers.

Furthermore, the training loss is also optimized with the fact that there is a specific image desired when starting with the conditional image. This training loss determines what images the generator strives to produce. Using solely the adversarial loss means the generator only seeks to fool the discriminator. This loss results in realistic textures but may not result in the desired image transformation. Pix2pix combines this loss with L1 loss, a loss that calculates pixel differences between the generated output and the desired output. The L1 loss ensures the output has similar structure to the desired image.

Since L1 loss covers the overall structure, the adversarial loss from the discriminator needs to focus on local textures. The discriminator architecture named PatchGAN is designed to output realistic textures at a higher resolution (Isola et al. 2017). This architecture divides the input image into smaller square patches of data. Each patch is separately evaluated as real or fake. The final classification is the average classification over all the patches.

Pix2pix only works with supervised learning, where training data between the conditional input and output image exist. An unsupervised model would be able to perform image-to-image translation between two previously unlinked domains. For example, with training data of cats and lions, the model could create photographic images of cats and lions in the same pose.

One architecture that implements unsupervised image-to-image translation is called UNIT (Liu et al. 2017). This architecture makes the assumption that a pair of images from two different datasets can be encoded into the same features. From the same features, the images can each be decoded back into their original form.

The UNIT architecture uses a pair of GANs, each one representing a different domain. The pair of GANs have shared encoder weights. Therefore, each GAN encodes into similar representations. These GANs are trained with the same encoder but different decoders. In order to translate between the two domains, an image is first transformed into features with the same encoder, and the decoder from the other domain creates an image out of these features.

Vid2vid is a GAN architecture which adapts the technology to produce a realistic and coherent video. Simply connecting generated image together would produce harsh frame transitions and inconsistency. To produce a realistic video, the model must consider neighboring frames during training. Two different discriminators are used to incorporate this information. One discriminator judges frames individually, ensuring each frame looks real. The other discriminator judges pairs of frames, focusing on the similarity and movement between frames. The adversarial loss is combined with optical flow, which measures how smooth the frame transitions are. Vid2vid is able to produce realistic frames that match surrounding frames and generates a smooth rendering of video.

3.2.2.2 *Face-specific synthesis*

Specialty GAN architectures are more capable in generating images within their domain. The following architectures are designed to work well on facial generation and manipulation.

StyleGAN demonstrates significant abilities for image synthesis on a dataset of face images (Karras et al. 2019). It uses training data from a high-quality dataset of real faces. StyleGAN utilizes progressive growing, where the model adds more resolution to the input and output after successfully training at a lower resolution. This technique results in large and clear output images. The architecture also learns styles and facial features in an unsupervised manner. Using this feature allows the model to control certain features in the output face, such as glasses, skin tone, and hair color.

Disentangled Representation learning Generative Adversarial Network (DR-GAN) combines the task of learning a representation of a face as well as generating any pose of the learned face (Tran et al. 2017). The generator uses an encoder–decoder architecture. The encoder takes in a facial image and transforms it into an encoding. The decoder takes in the encoding along with pose information to synthesize the face in a matching pose. The discriminator judges the generated image against the real image of the correct pose. Using this adversarial loss, DR-GAN can create a superior facial image with control over the facial pose.

DeepFake is an algorithm that replaces one face in a video with another face (Rossler et al. 2019). The term DeepFake now commonly refers to many face swapping techniques with varying architectures. However, the original method is significantly based on unsupervised image-to-image translation, or UNIT. The two different faces serve as the two domains. The shared encoder learns how to encode the features of the two different faces. The face is generated by using the shared encoder and generating an image with the other face's decoder.

3.2.3 Other face swap techniques

Video Face Replacement is one of the first automatic face swap algorithms (Dale et al. 2011). It utilizes a 3D modeling system to track a source face and target face from a single camera video. The source face is cropped and pasted over the target face. It is then warped to fit and match the 3D geometry. This technique carefully considers differences in alignment and size. In order to match the two faces together, the algorithm also calculates the best seam between the two faces to blend the border.

Face2face also uses 3D models and adds image rendering to blend in the source face to the environment and refines this technique (Thies et al. 2016). The algorithm adds motion and expression transfer to further match the source and target face. Furthermore, it is efficient enough to manipulate a video in real time, such as a live webcam feed.

These 3D modeling methods are efficient and fast as they transfer faces, but much post processing is needed to make the output look realistic. Furthermore, the quality of these videos is subjectively worse than deep learning-based methods.

3.3 DEEPFAKE USES AND THREATS

Algorithms collectively known as DeepFake were originally made for creating adult content where an actress's face is inserted into an adult video (Afchar et al. 2018). These methods were never directly published academically, but they exist in public repositories. These unethically produced videos caught the attention of mainstream media. The public reacted negatively to these AI-manipulated videos. While the original author is no longer active, community-developed implementations of the algorithm persist and are popular for creating videos for entertainment. DeepFake videos of movie scenes with all faces replaced with a single actor's face are popular.

3.3.1 Present threats

The threats of DeepFake videos extend past its original use. Another threat includes the dissemination of false information or fake news. For instance, multiple videos of Barack Obama exist, being manipulated through DeepFake technology. These videos have the appearance that President Obama is making outlandish statements, but these statements are false.

These fake videos can spread misinformation rapidly to the unsuspecting public. Multiple countries including the USA and China have passed laws addressing DeepFakes as a threat to national security. These laws limit the distribution of DeepFake videos without proper disclaimers. Social media and news websites are especially weak to the viral spread of fake videos.

These threats may also be used in organized crime and scams. Trusted professionals such as doctors or lawyers are hosting an increasing number of virtual meetings to connect with the public. AI-based professionals would also be possible in the near future. These digital appearances may be imitated by DeepFakes. An adversary may make false statements or gain access to private information.

3.3.2 Future robotics and DeepFakes

In the future, robotic systems are likely to integrate video feeds into their systems for various purposes. These tasks may include video conferencing between professionals and clients as well as presenting information digitally. One such robotic system can include telepresence systems. A telepresence system is designed to replicate a person's presence through a video

feed attached to a moving robot. These systems are used for telecommuters as well as employees physically distant from a meeting. As technology improves and work becomes more digitized, these systems may be integral to a working environment. A simpler version of telepresence includes virtual meetings, which are becoming increasingly common today due to COVID-19. Robotic systems may also include AI-based professionals such as doctors or lawyers. These systems may be able to provide advice and assistance through video. The AI-based professional would have a synthetic appearance to interact with clients.

Future robotic systems are vulnerable to DeepFake-based attacks as well. These digital appearances may be recreated by DeepFakes. An adversary may make false statements or gain access to private information. Video conferencing with trusted people such as doctors and lawyers can be breached since those trusted people may simply be images created by DeepFake. Furthermore, any AI-based professional in the future can have their appearance wrongly synthesized through DeepFakes. These attacks can have long-lasting effects on the public's trust in digital media, while digital media are likely to be an integral part of future robotic systems. Automated detection methods combined with cybersecurity technologies are needed to combat these threats.

3.4 DEEPFAKE DATABASES FOR RESEARCH

The advancements in deep learning has made video manipulation and synthesis much more accessible than before. Publicly available tools allow users to create their own forgeries and distribute them. Researchers are especially concerned when these tools are used to change the facial expressions or identity in a video and presented as a real video. Since these manipulated faces are not immediately discernible to a human observer, detection methods must be developed to identify these fake videos.

3.4.1 Research datasets

In order to compare different detection methods, a consistent set of data must be used. However, the availability of forged videos remains limited and hard to collect. FaceForensics++ aims to provide the large dataset of forged videos with 1.6 million images from more than 1,000 videos (Rossler et al. 2019). These videos are manipulated by four different facial manipulation techniques including DeepFake and Face2Face. Furthermore, the authors of the dataset include a public benchmark to test detection solutions against other algorithms.

Celeb-DF is another dataset aimed at addressing DeepFakes (Li et al. 2020). This dataset seeks to provide higher quality data to work on. Other,

older datasets have a noticeable gap in quality from DeepFakes that are circulated online. They also include poorly made forgeries that would not fool human observers.

To address these concerns, Celeb-DF focuses on the performance of the DeepFake generation model. It increases the resolution of generated faces to 256 by 256 pixels. The larger generated face resulted in less distortion and blur when resizing and warping to the target face. Next, the generated face is color corrected to match the skin tone and lighting of the original video. Lastly, placement of facial features are smoothed between frames, making the movement appear more natural.

The efficacy of these datasets depends on how hard it is to accurately classify the videos. More challenging data provide more research value to defend against stronger attacks in the wild. In one comparison with popular detection methods, Celeb-DF was proven to be the most challenging dataset to detect fake videos. However, as DeepFake methods improve over time, it is necessary to update research datasets with newer techniques.

3.4.2 DeepFake detection challenge

The DeepFake Detection Challenge (DFDC) is a contest designed to encourage research into detecting DeepFakes (Dolhansky et al. 2019). It is a joint effort between industry and academia with a total prize pool of $1,000,000. The DFDC was announced on September 2019 and ended in March 2020. It is hosted on the website Kaggle, which provides the training and testing datasets as well as a platform to run deep learning models on.

A notable contribution of this contest is the immense dataset of DeepFake videos that it provides. The DFDC dataset consists of more than 100,000 videos from a variety of paid and consenting actors. Other datasets such as FaceForensics++ collect and utilize videos from public sources such as YouTube. This use of actors addresses any ethical or legal concerns of using people's likeliness for research purposes.

This dataset also seeks to emulate real-world conditions where DeepFakes exist. The videos in the training dataset vary greatly in many different areas, similar to organically produced videos. The dataset accounts for differences in recording devices, resolution, brightness, contrast, and frame rate. Furthermore, the videos in the dataset have different compression rates, making it more difficult to examine details at the pixel level. DeepFake generated faces often contain artifacts and noise from the generation process that make it distinct from the rest of the image. Compression and resizing means that these differences are less noticeable, and the added noise is applied to real and fake videos alike. These details force a potential solution to focus on general features that can be applied to any DeepFake algorithm.

3.5 DEEPFAKE DETECTION METHODS

Detection methods for forged videos often use a convolutional neural network (CNN) approach due to their state-of-the-art performance in image recognition tasks. There are numerous proposed models aimed at detecting manipulated images. This section contains selected recent methods for forgery detection. The important features of a detection method include classification performance and generalization. Having high generalization means that the detection method can robustly detect different DeepFakes.

Early detection methods were trained on limited datasets. While they often show promising performance and high accuracy, these models significantly worsen on detecting DeepFakes from an unseen dataset. In a time where DeepFake methods are constantly evolving, generalization is important to create a meaningful defense.

3.5.1 General methods

MesoNet uses a CNN-based classification model to detect DeepFake and Face2Face manipulations (Afchar et al. 2018). The authors of MesoNet designed it to analyze the medium details in the image. Small, pixel-level features of an image can be affected by compression. High-level features such as face shape and structure will be generally correct after swapping faces. Therefore, medium-level features offer the best level to distinguish between real and fake images. In order to utilize this information, the architecture utilizes a smaller number of layers within the CNN. This model is advantageous in its small size and low overhead compared to other detection methods.

XceptionNet improves upon CNN architecture with depth-wise separable convolutional layers (Chollet 2017). XceptionNet was designed for general image classification on the ImageNet dataset. Due to its outstanding results, it was adapted to DeepFake detection through transfer learning. The final layer designed to classify the image into 1,000 different classes is replaced with a layer designed to classify between real and fake. This model is then trained on the FaceForensics++ dataset (Rossler et al. 2019). The video in question is preprocessed into individual frames and ran through the model.

EfficientNets are a newer family of CNNs that are designed for efficient scaling (Tan and Le 2019). By balancing the scaling of model resolution, width, and depth, EfficientNets outperforms models of comparable parameters. There are eight different models varying in size, labeled from B0 to B7. These models show promising results on classifying general images, so they can be useful on classifying DeepFakes.

ForensicTransfer utilizes an autoencoder-based architecture to detect DeepFakes (Cozzolino et al. 2018). The autoencoder consists of an encoder–decoder pair. It learns the feature representation of an image.

It then classifies it as fake if it is sufficiently distant from the cluster of real images. This solution is designed to improve generalization or the ability to work on different datasets. ForensicTransfer is comparable to other methods when working on the same dataset for training and testing. Its performance is noticeably better when testing on different datasets. This ability can be useful when detecting images manipulated through an unknown method.

3.5.2 Specific features

A model with long-term recurrent convolutional network (LRCN) can target certain weaknesses within the video movement and behavior. A recurrent neural network such as LRCN considers previous frames as well as the current frame to generate output. The final output of the model is based on each frame of the entire video, allowing it to examine behavior over time.

For instance, early versions of DeepFakes failed to account for the human behavior of blinking. Using cropped images of the eyes in a video, a model that utilizes LRCN can be trained to detect the number of times the subject blinks in the video (Li et al. 2018). Next, the model can compare the detected blinking to an estimated rate of 17 blinks per minute. The model decides if the video is real or fake based on this comparison.

Other models utilizing recurrent neural networks examine the general temporal consistency within the video (Güera and Delp 2018). The frame-to-frame transition of generated faces may have artifacts from the generation process. Examining these with an appropriate model can differentiate real and fake videos.

Head Pose is a model that examines inconsistencies in 3D head poses within videos (Yang et al. 2019). The model crops and detects facial landmarks within the video. Next, the model estimates the 3D pose of the facial landmarks. In poorly made DeepFakes, the 3D pose of the face is not aligned correctly with the rest of the head. The estimated 3D pose is fed into a classification model to determine if the video is real or fake.

3.6 IMAGE SEGMENTATION

Another area of research interest involves identifying manipulated regions through image segmentation models. Image segmentation involves assigning a label to each image pixel based on the surrounding area. Applying this technique to DeepFakes identifies manipulated faces within the videos.

One popular architecture for segmentation is U-Net (Ronneberger et al. 2015). Earlier solutions relied on a sliding window approach, where the image was split up into multiple overlapping patches. The model would then classify the center area of the patch based on the surrounding area. However, this approach was costly and slow, since each patch had to go

through the model, increasing redundancy. U-Net addresses this problem with matching the downsampling encoder with an upsampling decoder. This decoder increases the area covered by the output, removing the need to have sliding windows. The context is preserved with increased feature maps and concatenation from the corresponding encoding layer. The concatenation is also called skip connections. Information skips the bottleneck of going through the entire encoder, which reduces the resolution in favor of more features. By connecting these layers, the decoder can utilize structural information of the image when upsampling the features. This architecture creates a U shape with the encoder descending on the left and the decoder ascending on the right.

Y-Net extends the U-Net architecture with a classification branch (Mehta et al. 2018). This combined architecture was originally designed to simultaneously segment and classify breast biopsy images. The U-Net portion of the architecture segments the image into cells of different types. A classification branch is added to the end of the encoder, forming the bottom of the Y-shaped architecture. This branch is trained to diagnose the image as benign or malignant based on the encoding. This combined model is useful for solving multiple related tasks.

These segmentation models are originally designed for biomedical segmentation. However, they can be adapted to the fields of DeepFake detection with the appropriate training data and careful adjustments. The segmentation model needs existing pairs of images that contain fake faces and masks illustrating the altered region. These masks can be created through examining the pixel differences between the original and fake videos. They may also be created during the generation of the fake face, illustrating the part of the image the fake face is inserted into.

3.7 MULTI-TASK LEARNING FOR DEEPFAKES

Incorporating a segmentation model into a detection method has been shown to improve performance. Information gained through the segmentation branch and classification branch is shared throughout the model, improving the performance of both branches. One multi-task learning approach combines three different tasks to identify and segment manipulated faces in video (Nguyen et al. 2019). The multi-task model is designed to output a classification, segmentation, and reconstruction of the image. The loss for all three of these tasks are summed up with equal weights to form the total loss. This model is trained to minimize the total loss, so it will learn all three tasks simultaneously.

Multi-Task learning has promising results, but more research is needed to advance the model. The existing model uses a simple segmentation and classification branch, leaving room for improvement. Using a more advanced segmentation model would improve results. One proposed model

would combine a segmentation model that utilizes U-Net with a classification branch. These two tasks are related for detecting DeepFakes and altered digital media. Using an advanced CNN within the U-Net could strengthen the model as well. EfficientNets are state-of-the-art and can be adapted for this use.

3.8 CONCLUSIONS AND FUTURE RESEARCH

3.8.1 Summary

DeepFakes are a collection of algorithms that use deep learning to change faces within videos. This chapter first described the primary technology used within DeepFakes; i.e., GANs, including how conditional GANs allow more control over the output and are useful for image synthesis, and how numerous methods improve upon this technology to generate realistic images and faces. Next, the chapter illustrated DeepFakes which utilizes these techniques for face swapping algorithms. We also explained how future robotic systems are likely to include video feeds within the workplace and video representation of AI-based professionals, and that these systems are vulnerable to DeepFake manipulation as videos can be easily manipulated. In order to address these vulnerabilities, the chapter described in detail DeepFake datasets and detection methods. We noted that while some detection methods showed good results, further research is necessary to secure robotic systems against the threat of DeepFakes and manipulated video.

3.8.2 Concluding remarks

Facial manipulation methods utilize traditional 3D modeling and computer graphical methods to change the facial identity of people in videos. Deep learning models commonly known as DeepFakes can also alter facial images to a high degree of accuracy. This technology may possibly be used for malicious intent and deception. DeepFakes are a growing concern for digital media and future robotic systems. The high-quality appearance of the output makes it difficult to distinguish from reality. These video manipulation tools are widely available and are continuing to improve.

These tools pose many threats to spread disinformation, erode trust in digital media, and weaken national security. Classification models exist for known manipulation methods, but these models are vulnerable to newer attacks. Models need to detect a wide range of facial manipulation techniques to implement effective detection. Benchmarks and challenges such as the DFDC promote competition and stimulate research into this field.

Current manipulation techniques are difficult for humans to identify, so automated detection methods are needed to identify fake videos.

As described in this chapter, CNN-based classification models are able to identify known facial manipulation with reasonable accuracy, but they are still too weak at detecting unseen methods, and require rigorous research to improve detection of facial image and video manipulations.

3.8.3 Future research directions

More research is needed to create a robust DeepFake detection model to secure future robotic systems. One possible area of interest for improving detection systems is examining the visual difference between the face and background using image segmentation. Techniques based on these differences would be able to generalize well on diverse manipulation techniques. Combining the task of image segmentation and classification improves the performance of the model. Furthermore, DeepFake videos may be able to reduce the differences between generated and real faces. Detection methods must be able to adapt and improve along with the generation methods.

Research into addressing DeepFakes within robotics must also consider evolving technologies. Robotic systems in the future may incorporate video feeds for conferencing and telepresence systems. To ensure security and reliability, video systems need to automatically prevent attacks from malicious actors who could misrepresent themselves on video. Furthermore, AI-based professionals can interact with the public through video as well. These systems need strong authentication to ensure that the video and information is from the correct AI sources. Reliable and accurate DeepFake detection methods are integral to the security of future robotics and digital media.

REFERENCES

Afchar, D., V. Nozick, J. Yamagishi, and I. Echizen. 2018. Mesonet: A compact facial video forgery detection network. *Paper read at 2018 IEEE International Workshop on Information Forensics and Security (WIFS).*

Chollet, F. 2017. Xception: Deep learning with depthwise separable convolutions. *Paper read at Proceedings of the IEEE Conference on Computer Vision and Pattern Recognition.*

Cozzolino, D., J. Thies, A. Rössler, C. Riess, M. Nießner, and L. Verdoliva. 2018. ForensicTransfer: Weakly-supervised domain adaptation for forgery detection. *ArXiv abs/1812.02510.*

Dale, K., K. Sunkavalli, M. K. Johnson, D. Vlasic, W. Matusik, and H. Pfister. 2011. Video face replacement. *Paper read at Proceedings of the 2011 SIGGRAPH Asia Conference.*

Dolhansky, B., R. Howes, B. Pflaum, N. Baram, and C. C. Ferrer. 2019. The Deepfake Detection Challenge (DFDC) Preview Dataset. *arXiv preprint arXiv:1910.08854.*

Goodfellow, I., J. Pouget-Abadie, M. Mirza, et al. 2014. Generative adversarial nets. *Paper read at Advances in Neural Information Processing Systems.*

Güera, D., and E. J. Delp. 2018. Deepfake video detection using recurrent neural networks. *Paper read at 2018 15th IEEE International Conference on Advanced Video and Signal Based Surveillance (AVSS).*

Isola, P., J.-Y. Zhu, T. Zhou, and A. A. Efros. 2017. Image-to-image translation with conditional adversarial networks. *Paper read at Proceedings of the IEEE Conference on Computer Vision and Pattern Recognition.*

Karras, T., S. Laine, and T. Aila. 2019. A style-based generator architecture for generative adversarial networks. *2019 IEEE/CVF Conference on Computer Vision and Pattern Recognition (CVPR)*:4396–4405.

Li, Y., M.-C. Chang, and S. Lyu. 2018. In ICTU oculi: Exposing AI created fake videos by detecting eye blinking. *Paper read at 2018 IEEE International Workshop on Information Forensics and Security (WIFS).*

Li, Y., X. Yang, P. Sun, H. Qi, and S. Lyu. 2020. Celeb-DF: A large-scale challenging dataset for DeepFake forensics. *Paper read at Proceedings of the IEEE/CVF Conference on Computer Vision and Pattern Recognition.*

Liu, M.-Y., T. Breuel, and J. Kautz. 2017. Unsupervised image-to-image translation networks. *Paper read at Advances in Neural Information Processing Systems.*

Mehta, S., E. Mercan, J. Bartlett, D. Weaver, J. G. Elmore, and L. Shapiro. 2018. Y-Net: Joint segmentation and classification for diagnosis of breast biopsy images. *Paper read at International Conference on Medical Image Computing and Computer-Assisted Intervention.*

Mirza, M., and S. Osindero. 2014. Conditional generative adversarial nets. *arXiv preprint arXiv:1411.1784.*

Nguyen, H. H., F. Fang, J. Yamagishi, and I. Echizen. 2019. Multi-task learning for detecting and segmenting manipulated facial images and videos. *arXiv preprint arXiv*:1906.06876.

Ronneberger, O., P. Fischer, and T. Brox. 2015. U-net: Convolutional networks for biomedical image segmentation. *Paper read at International Conference on Medical Image Computing and Computer-Assisted Intervention.*

Rossler, A., D. Cozzolino, L. Verdoliva, C. Riess, J. Thies, and M. Nießner. 2019. Faceforensics++: Learning to detect manipulated facial images. *Paper read at Proceedings of the IEEE International Conference on Computer Vision.*

Tan, M., and Q. V. Le. 2019. EfficientNet: Rethinking model scaling for convolutional neural networks. *Paper read at ICML.*

Thies, J., M. Zollhofer, M. Stamminger, C. Theobalt, and M. Nießner. 2016. Face2face: Real-time face capture and reenactment of RGB videos. *Paper read at Proceedings of the IEEE Conference on Computer Vision and Pattern Recognition.*

Tran, L., X. Yin, and X. Liu. 2017. Disentangled representation learning GAN for pose-invariant face recognition. *Paper read at Proceedings of the IEEE Conference on Computer Vision and Pattern Recognition.*

Yang, X., Y. Li, and S. Lyu. 2019. Exposing deep fakes using inconsistent head poses. *Paper read at ICASSP 2019-2019 IEEE International Conference on Acoustics, Speech and Signal Processing (ICASSP).*

Chapter 4

Nanoscale semiconductor devices for reliable robotic systems

Balwant Raj
Punjab University SS Giri Regional Centre Hoshiarpur

Jeetendra Singh
NIT Sikkim

Balwinder Raj
National Institute of Technical Teachers Training and Research

CONTENTS

4.1 INTRODUCTION

The technology advancement enables the device-related improvement such as low operating power, high speed, and high performance for sensors. The metal oxide field effect transistor (MOSFET) structure dimensions such as gate length, device width, and dielectric thickness under gate have been downscaled to nanoscale dimensions with passing years [1–5]. Downscaling has the advantage in terms of low power supply requirement, minimum delay, and more packaging density. Microelectronic devices will play a key role in the future of nanoscale sensor design for environmental applications. Traditional microelectronic devices such as complementary metal oxide semiconductor (CMOS) are scaled into the nanoscale dimensions and are developing exceptionally inexpensive but unexpectedly controlling devices and circuits [7]. Also, the current CMOS devices and circuits process technology further to be useful as a foundation to design new nanoelectronic devices for future applications. It is not unlikely that a new design combining microelectronic and nanoelectronic devices develops advanced device circuit co-design system-level applications [7–9].

The nanoscale semiconductor devices with some modifications in traditional CMOS transistors such as change in material, change in device structure, increase in number of gates instead of one gate, metal gate (MG) MOSFET, and FinFET devices are also becoming popular [10–16]. The knowledge of working mode and the restrictions of conventional MOS transistors are required to understand these advanced device engineering. Scaling of conventional MOS devices into the nanometer regime requires additional device materials and architectures. Suitable nano devices are silicon on insulator (SOI) with more gates, e.g., double-gate (DG) FinFET structures, that can be scaled down in nano dimensions than traditional bulk silicon devices [17–20].

The main importance of multi-gate MOSFET devices is more effective electrostatic control of charge carriers under gate area, suppressing effectively the short channel effects (SCEs). The quantum mechanical (QM) effects play a key role in these nanoscale devices with a very thin gate dielectric material and Si body. Scaling down conventional bulk CMOS transistors (sub-50 nm) is very difficult due to the high SCEs and leakage currents [21]. To reduce the leakage power, dielectric materials under the gate required high dielectric constant, and substrates were doped largely. DG FinFET devices mitigate that SCE problems occur due to downscaling in the nano domain [22–24]. The leakage current is reduced by adding another gate on the back side, which doubles the channel to gate capacitance and provides better controllability over the device channel potential [25].

With key structural benefits of fully depleted semiconductor, FinFETs are low power devices because of insusceptibility of SCEs and leakage currents [26]. FinFET structures with DG having size of 10 nm are studied in the

literature [10]. As compare to conventional MOSFETs, FinFET structures needed two-dimensional (2-D) simulations to analyze their detailed current behavior [27]. But with scaling down parameters in the nano domain, a classical device study is not sufficient. A detailed study of Schrödinger–Poisson equation is required to precisely analyze the FinFET behavior. The simulation studies are extremely required due to huge grid points in 2-D problems. They are basically not suitable. In place of this, classical device simulation study with additional QM models can be developed. However, the comparison and validity of these newly developed models for ultra-thin silicon FinFET devices are currently in progress. FinFETs were developed due to their process simplicity and compatibility with current process flow for bulk planar MOSFETs. The self-alignment for the FinFET structure makes it inherently advantageous as compared to bulk CMOS technology or standard DG MOSFET because of lower values of intrinsic gate to source and gate to drain capacitances, which in turn result in high speed of operation. Furthermore, by varying nature of structures of the FinFET, there is a reduction in the leakage current and hence an increase in I_{on}/I_{off}. The progression of the MOSFET device size scaling is kept on fast track due to rapid development of process techniques. As MOSFET device is scaled down, SCEs increase and leakage currents are enhanced [28,29].

DG FinFET was proposed to mitigate the SCEs in the future nanoscale device. For very short channel length devices within the small volume of the Si channel, a less deviation in the number of doped atoms will have a large effects on the effective doping profile. Hence, as per the classical study of relationship between the V_{th} and doping profile, controlling V_{th} very precisely will be a challenging task and is likely to become a serious issue due to doping profile variation. Although some reported papers [30] had discussed this problem, this mainly reports the conventional method of high doping for controlling V_{th} variation. To extend the downscaling limits for nanoscale technologies, advanced device structures of fully depleted (FD) SOI and DG FinFET [11,31] with undoped or very less doped ultra-thin devices are emerging [32]. The increase in V_{th} has been obtained in the thin body semiconductor device. The reported works analyzed the V_{th} shift numerically [33] and preserved the QM effect with the effect of a high impurity concentration. An analytical model is required for understanding the reported experimental data and provides guidance to the V_{th} adjustment for ultra-thin body field effect transistor.

Given such a promising importance, device modeling of FinFET has become a necessity for possible future applications. The leakage currents associated with bulk MOSFET and DG FinFET have been described in detail [34], while the circuit aspects of FinFET-based static RAM (SRAM) have been studied by Raj et al. [2,35,36]. They showed that because of excellent control of the SCE in FinFET, there is a lower subthreshold leakage current, and therefore, FinFET developed as one of the key promising devices for advanced circuit design in the nanoscale regime.

In this chapter, the review of the existing literature is presented. Various research papers, books, and monographs are referred to which take care of various aspects such as: classical and QM modeling of FinFET devices in order to understand the suitability of the work being carried out as well as to understand the various technical gaps in the area of FinFET. Section 4.2 covers the literature on scaling of bulk MOSFET devices and Section 4.3 deals with a survey of various advanced MOSFET structures, inversion charge density, and drain current. Furthermore, in this chapter, Section 4.4 describes the review of FinFET devices, and Section 4.5 enumerates the research performed on leakage currents associated with the FinFET device. The sensor aspect of the FinFET device has also been presented, and this chapter concludes with Section 4.6.

4.2 BULK METAL OXIDE FIELD EFFECT TRANSISTOR (MOSFET) SCALING TREND

Scaling of MOSFET without changing the inside electric field of device gains special interest and in the literature, many articles published, which have become very famous in the semiconductor device community. Authors reported that all semiconductor device dimensions to be scaled down by $1/\kappa$, whereas the doping of the source side and drain side should increase by a multiple of κ. Applied supplied voltages should be scaled down by $1/\kappa$ [37]. The planar bulk-silicon MOSFET shown in Figure 4.1 has been the well-known of the semiconductor industry from the last four decades. In 1980s, many alternative models to bulk silicon MOSFETs have been proposed. However, the scaling down of conventional MOSFET becomes gradually hard for channel lengths below ~10 nm, predictable by the year 2020. On reducing the channel length, it is observed that the capacitive coupling between channel potential source/drain (S/D) was higher as

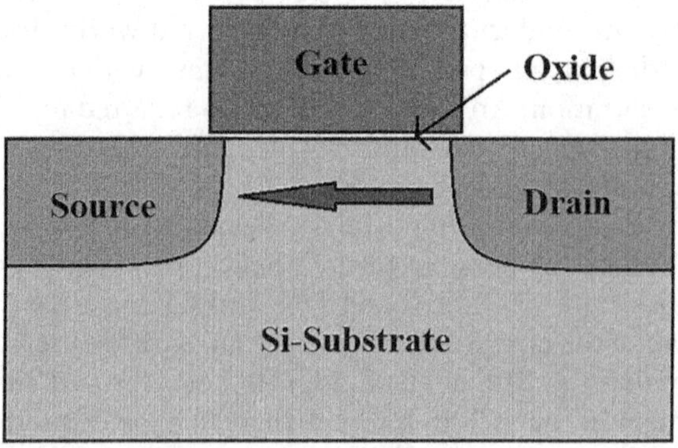

Figure 4.1 Bulk MOSFET structure of device required for circuit design.

compare to the gate, which causes poor SCEs. The key merit for minimizing the dimensions of the device has increased transistors density per unit chip area, improved speed of operation, and reduced cost per function. Device scaling requires a balance between device functionality with device reliability. Both of these characteristics have to be maintained as one scales channel length to smaller sizes. As critical transistor dimensions are scaled, reliability concerns have become more pronounced. This manifests itself as [38–42]: (i) increased OFF-state current, (ii) threshold voltage (V_{th}) roll-off, (iii) drain-induced barrier lowering (DIBL), which is basically modulation of the source-channel potential barrier by the drain voltage and arose due to reduction of V_{th} with increasing drain bias.

In order to build the relatively good gate control of the channel potential in conventional MOS devices, many developments such as ultra-thin gate dielectric materials [43], ultra-shallow S/D junctions [44], new doping process, i.e., halo implants [14,15] and super-steep retrograde wells, known as improved channel impurities profile engineering techniques have been essential. Although, these technologies have become most promising in fundamental physical limits which saturates further downscaling of semiconductor device dimensions [45]. In order to enable device downscaling of conventional MOS devices, the gate dielectric material thickness is found to be the single attractive device dimension to reduce. The capacitive coupling between gate and channel can be improved by a thin gate dielectric material, which overcomes the influence of the S/D on the channel.

A large gate capacitance is capable to develop large inversion charge density, which improves the ON-state current of the device. However, there is QM direct tunneling that occurs through thin gate dielectrics, which contributes the substantial gate leakage currents below ~20 Å. This gate leakage can be reduced by the usage of other high-k gate dielectric materials, high k gate dielectric constitutes a small effective oxide thickness and builds a large barrier for QM direct tunneling, and thus provides better gate control required for L_g scaling [46]. The leakage current can also be reduced by acquiring a heavy doping in the body of the bulk Bulk-Si MOS transistor and increasing the back-gate control of the body, thereby more scaling of the Bulk MOS transistor can be done. In the case of sub-50 nm channel length devices, generally a strong halo implant is preferred to overcome sub-surface leakage, however this method raises the mean doping of the channel small L_g devices. However, a large doping concentration in the channel leads to the impurity scattering, which causes less carrier mobility, and also subthreshold slope (SS) will be raised due to high transverse electric field, along with high band-to-band tunneling leakage and large depletion and junction capacitances. The device performance is collectively degraded by these factors significantly [1,5,8,12].

In summary, according to the semiconductor device designer standpoint, the gate dielectric oxide thickness (T_{OX}), depth of the S/D junction (X_J), depletion width of channel (X_{DEP}) need to be scaled down so that a good

electrostatic control of SCE could be achieved. l_{BULK}, basically specifying the length of L_G is to be scaled down so that SCEs do not become destructive. For a conventional MOS transistor, T_{OX} downscaling is limited by gate leakage, and scaling of X_{DEP} is restricted to about 10 nm. The body doping is restricted by substrate-to-drain and band-to-band tunneling current. Whereas, ultra-shallow junctions containing abrupt doping profiles limit X_J [47]. Moore's device scaling law suggests that the number of MOS transistors on integrated circuits (ICs) approximately doubles in every 18th months. The decrease in minimum feature size in ICs can place more devices in small area, hence chip density increases. Its leads to small power consumption and increases the performance. However, the least feature has decreased, the benefits of technology downscaling have been reduced today. One of the barriers of technology downscaling is related to power dissipation and consumption. In day-to-day life, we see that the power consumption reduces and hence the performance is analyzed at the cost of power dissipation and consumption. Supply voltage has remained almost constant, but leakage current increased exponentially. In the design of radiation hardened memory, we should also take care about the voltage scaling.

4.2.1 Challenges of bulk complementary metal oxide semiconductor (CMOS)

In order to achieve performance in terms of high speed and low power operation to raise the density of MOS transistors in chip, Moore's device scaling law and the market are taken as key references. In the case of planar transistor with the length below 20 nm, the device scaling process now saturates or seems to be at a technology edge. The bulk CMOS technology faces other challenges as dimensions are scaled down. In order to keep SCEs under control, ultrashallow junctions with very high doping abruptness and yet high degree of dopant activation are required. Although, some methods such as laser annealing and flash lamp annealing are currently being investigated, these may not work for future technology nodes [48]. In addition, the poly-Si gate depletion effect contributes significantly towards the effective oxide thickness and hence the V_{th} and performance. This effect can be completely eliminated by moving back to the metal gate technology. negative-channel metal-oxide semiconductor (NMOS) and positive channel metal oxide semiconductor (PMOS) devices, however, need separate gate materials to achieve the required work functions, leading to process integration challenges. Even though bulk CMOS technology with SiO_2 gate dielectric and poly-Si gates has been the most suitable and well-understood technology, the abovementioned challenges strongly push the need for alternate device structures and processing techniques [49–52]. Static power consumption arises because SCEs are found to be the key challenge posed by the limits of downscaling [6,53] and the leakage

currents [54,55]. The leakage currents are also involved due to quantum mechanical tunneling in semiconductor devices, the subthreshold leakage current, junction leakage current, direct tunneling between S/D because of the channel potential barrier. Replacing the conventional dielectric material SiO_2 by other materials with high-k dielectric constant, it will be used to suppress some of leakage current problems in semiconductor devices. The subthreshold leakage current is not blocked by introducing alternating materials. High-K materials are proving to be one of the crucial limits for scaling the size of semiconductor devices. In addition, the polysilicon gate electrode terminal has the restriction related to the depletion area and boron out-diffusion, decreasing valuable effects of device downscaling. As a result, other metal gate materials will be required in combination with high-k dielectric materials. Advancements in both the semiconductor device design and the various materials will confirm high speed action of nanoscale semiconductor devices. Some of the additional auspicious methods utilize SOI, strained silicon-germanium, high-k electrode, metal gate electrode, FinFET, DG MOSFETs, nanowire, and carbon nanotube field-effect transistor (CNTFET).

4.2.2 Limitations of bulk MOSFET

The CMOS devices based on SOI have many advantages over Si base bulk CMOS devices from theoretical point of view [16,56] For example, in bulk silicon device contend with: (i) S/D to body or S/D to isolation of oxides which make parasitic capacitance; (ii) if device is continuously scaling, the SCE is shown; and (iii) degrading performance of the device.

The electrical factors degrading and the silicon process parameter variations affect the performances which is the reason of this planar device structure in which the gate does not have a enough electrostatic field to control the channel. Due to scaling down the device, it brings to:

i. Less current in the channel of device
ii. Leakage current of device drain-source when the MOS transistor is switched off
iii. SCEs
iv. High dependence on process variations (V_{th} and swing)

4.2.3 Design problems of nanoscale CMOS

i. CMOS latch up,
ii. Hot carrier gate dielectric degradation,
iii. Punch through
iv. Degradation in carrier mobility,
v. Source/drain series resistance
vi. Discrete doping effect.

CMOS Latch up: the conventional CMOS structure is made of two parasitic bipolar transistors. The Bipolar Junction Transistor (BJT) collector was connected to the base of the other transistor in a positive feedback configuration. When both transistors conduct, low resistance path is created between power supplies and ground. This process is called as latch up. It also occurs when two transistors gain product which is greater than one in the feedback loop. The result of this latch up problem is the destruction of the semiconductor device.

Hot Carrier Gate Dielectric Degradation: if the electric field is greater enough near the Si–SiO_2 interface, a region of high electric field is created. Due to this electric field the charge carriers gain enough energy and enter in SiO_2 layers. In general, the hot electrons inject more as compare to hole; there are two main reasons: (i) Electrons move faster than holes due to their less effective mass. (ii) The Si–SiO_2 interface energy barrier is higher for holes (4.5 eV) than for electrons (3.1 eV). This effect is called hot carrier injection in semiconductor devices. The hot carrier effect is also a main source of power dissipation in nanoscale semiconductor devices.

Punch Through: due to scaling the channel length, the depletion region, the drain-substrate, and substrate–source junction extend into the channel. When the increase in the reverse voltage across the junctions (with increase in V_{ds}), this also leads to the boundaries to be pressed extra away from the junction. When the device channel length and reverse bias give reason for the depletion layers to merge, then punch-through occurs. In sub-micron MOS transistors, better expansion of the depletion region under the surface gives rise to punch through.

Degradation in Carrier Mobility: As channel lengths shrink below 50 nm, MOS transistor processes require gate oxides thickness to be less than 1.5 nm. Due to this feature, the transverse electric field reaches much greater than 10^5 V/cm even when the semiconductor device is biased at threshold. This much greater electric field always leads to mobility degradation as scattering near the Si/SiO_2 increases. In addition, the high channel doping is used to lower SCE and adjust V_{th} results in mobility degradation due to Coulomb scattering with ionized dopants [13]. Although due to quantum effects, the inversion charge has a peak below the surface leading to the increase in oxide thickness of around 10 Å and decrease in the effective electric field, the degradation is large enough to decrease the drive current.

S/D Series Resistance: When MOS transistor is scaling, another preventive factor is affecting the external S/D resistance in junction of semiconductor devices. SCEs and source-drain series resistance are the fundamental challenges in current semiconductor devices [3]. For an MOS, the transistor works as a VLSI component that has the property to reduce leakage current and minimize the SCEs. However, when the channel resistance reduced as the side of S/D, there are no more improvements in drain current because increasing this resistance leads to overcome the SCEs. According to International Technology Roadmap for Semiconductors (ITRS) roadmap

predictions, S/D extension depths as narrow as 10 nm to attain a 50 nm MOS transistor, narrower junction leads to larger external resistance, which results in deterioration of driving current.

Discrete Doping Effect: the feature size of MOSFET is being scaled down to sub 50 nm, the total number of dopants implanted inside the channel becomes very less. The effect is that even a small fluctuation can give rise to significant differences in threshold voltage. The name discrete dopants comes from the fact that the total count of dopants comes only to few hundreds, and thus even a small number change appears to be significant, resulting in variation of V_{th}. There are only few problems that have a changing device performance and reliability. While original methods to scaling have improved some of these difficulties, it became clear that technological approaches are needed to push conventional CMOS further into the nanoscale regime. In 1980s, various methods to Si substrates gained momentum and described the devotion of the industry – Silicon-on-Insulator, known as SOI [57,58].

4.3 NANOSCALE MOSFET STRUCTURES

In order to mitigate some of the issues of the bulk MOSFET, following advanced transistor structures have been proposed:

- Ultra-Thin Body FET (UTB FET),
- DG-FET,
- SOI MOSFET.
- Multiple-Gate MOSFET (MG MOSFET).

4.3.1 Ultra-thin body (UTB) MOSFET

An UTB MOS transistor has been shown in Figure 4.2. The simple concept of the UTB MOS transistor is to use an extremely thin (<20 nm) SOI film to remove subsurface leakage current. The self-aligned S/D are required to

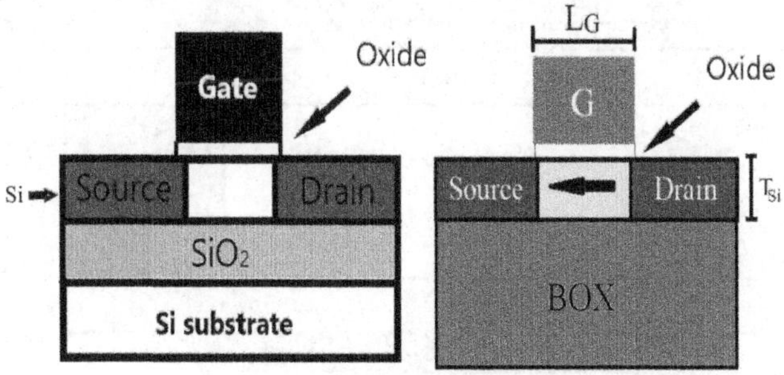

Figure 4.2 Ultra-thin body MOSFET structure having better control of shot channel effect (SCE) than conventional MOSFET.

diminish parasitic series resistance and achieve greater drive current. The key benefit of these structures is that the conduction is confined to a thin silicon film, thereby [59–61]:

i. Physically eliminating the sub-surface leakage component.
ii. The layouts and process steps are very close to the conventional bulk CMOS flow.
iii. An undoped channel is used to reduce the effect of statistical dopant fluctuations.
iv. They also have additional benefits of better short channel control.
v. Reduced parasitic capacitance (no source or drain-bulk capacitances exist).
vi. Overall, these devices show superior performance (I_{ON}/I_{OFF} as well as intrinsic delay) compared to their bulk counterpart [62,63].

4.3.2 Double-gate (DG) MOSET

The thin body essential can be stress-free, and device structure of DG MOSFET is shown in Figure 4.3, where top gate and bottom gate both control the potential of channels in this device. The DG-MOSFET attains better gate control and reduces the SCEs [64]. In the case of DG-MOSFET, the body thickness can be double as compare to a single-gated MOSFET, also realizes the similar degree of SCEs. The DG-MOSFET has no effect of electric field penetration on the S/D side, can be scaled in the nano domain. The thin body device extremely required from a design and development point of view, uniform ultra-thin film can stance major technological issues. Fabrication and simulation results reported in the literature indicate that a DG-MOSFET can be scaled down to below 30 nm channel length

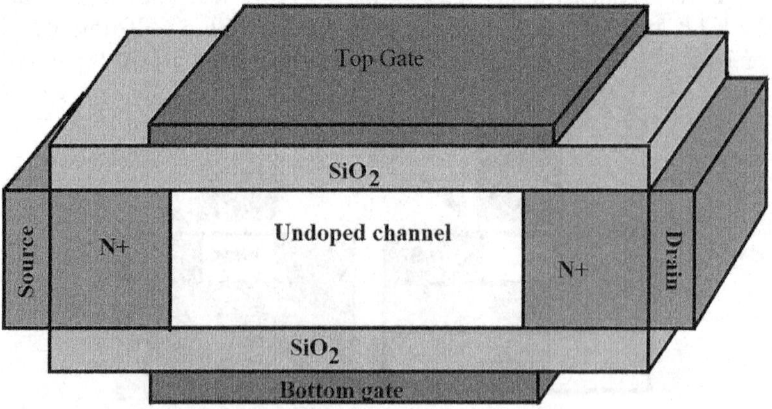

Figure 4.3 Double-gate (DG) MOSFET for high driving current and less leakage current for low-power circuit design.

(L_G) devices. The upgraded scalability of thin body semiconductor devices is suitable for future CMOS technology and so have fulfill the expectation of the ITRS [65].

The FinFET is DG MOSFET structure and can be scaled down these device below 25 nm regime. DG-MOSFET is designed as a very thin Si channel and is fully depleted which has negligible surface leakage current, that is possible by manufacturing these devices in the nano domain. The procedure of manufacturing that has less doping or undoped channels leads to improved resistance of the channel, smaller drain-to-body capacitance, and carrier mobility is very high due to lower effect of the electric field. If the channel is doping less, gates metal work function is essential to attain workable threshold voltages in fully depleted semiconductor devices [9]. The parasitic resistance and contact resistance side of S/D are very high with the thin Si channel that is a major issue. The parasitic resistances are the major problem in bulk semiconductor devices [66], the problem is very simple in thin body semiconductor devices, and different technologies can be used to reduce this problem. This chapter represents a device optimization technique to reduce the SCEs and thus external parasitic resistance. It compromises between the different device process parameters in defining the short-channel behavior. It can also be calculated by scaling the length of the channel. This is significant from the lookout of device scalability.

4.3.2.1 Advantages of double-gate MOSFET

i. Reduction of I_{off} that means leakage current will be very small
ii. Undoped channel removes intrinsic parameter variation and minimizes impurity scattering.
iii. DG MOSFET tolerates for higher current drive capability.
iv. It has good control of SCEs. It is possible because of two gate controllability over the channel.

4.3.3 Silicon on insulator (SOI) MOSFET

SOI devices are now under search for the development of advanced ICs with special properties: full isolation, low parasitic capacitances, vertical junctions, very high speed, lowest hot carrier booster, SCEs, simple device structures, and brilliant resistant to radiation effects. The most important feature of the SOI MOS transistors is dual-gate that controls the Si channel and electrostatic potential [67]. SOI MOSFETs are with bulk semiconductor devices to increase the speed of VLSI circuits and improve the frequency for CMOS applications [5,68]. As compared to planar bulk MOSFETs, the SOI MOSFETs are reduced SCEs because of their thin channel region. These devices are ideal semiconductor devices to allow further physical strictures downscaling of ICs.

As compare to bulk MOSFET, the SOI CMOS devices have reduced SCEs and have lesser power-delay product that leads these devices for a real application circuit design and also suppressed of latch up, better protection to radiation, higher speed, and less power dissipation. There are certain applications, which make the SOI more prevalent as compare to the bulk technology. As SOI microprocessors show an improvement of 20% in speed, SOI radio frequency (RF) power amplifiers have good power efficiency and low noise in the low-GHz range. Due to the ongoing scaling of the parameter of the bulk transistor technology, the numbers of metal layers have been increasing due to connection of the growing number of transistors on the chip. Because of this inevitable trend, interconnection is becoming more complex and thus interconnect delay is also a more dominant factor in the circuit speed. On the basis of researchers' standpoint, due to allowance of 3-D integration of devices with minimum interconnect complexity in SOI technology, it should develop very soon as the base of CMOS technology.

All these key features of the SOI MOSFET make the structure viable with Moore's Law far below the 0.1 μm node to the nanometer scale. The current evolving status of 3-D multiple-gate SOI MOSFETs clearly indicates that for high-performance applications of microprocessor and wireless, this new technology is frequently accepted as the mainstream of the CMOS technology. The transceivers, developed by the CMOS technology, are frequently taking on the wireless application, also SOI CMOS is making durable inroads into new evolving RF-based applications. Many critical difficulties still remain to be overcome, such as the gate dielectric development using high-K dielectric materials, lower dielectric materials for high speed interconnect isolation, uses of metallic materials in gates, and passive RF components. In spite of that, it is now evident that 3-D SOI MOS devices have become most promising means for possible nano MOSFET devices to improve Moore's Law predictions for the next generation.

4.3.4 Multiple-gate MOSFET

Multiple-gate MOSFET architectures of undoped SOI are showing potential structures possible to mitigate SCEs in advanced nanometer-size semiconductor devices. Similar to conventional MOSFETs, DG, Triple-Gate (TG), and Quadruple-Gate (QG) MOSFET need not require severe doping channel engineering. Furthermore, they permit relaxing the oxide thickness of device (T_{ox}) and the film thickness (T_{Si}) requirements that are difficult in FD conventional MOSFETs (SG) on SOI. The benefits of multi-gate structures are:

i. Higher drain current thus better performance
ii. Prophesized to show better tolerance to downscaling.
iii. Good integration feasibility, raised S/D structure, eases in devices integration at the circuit level.

iv. Large number of process parameters to modify device higher performance
v. Undoped structures, so no discrete dopant effect

The most promising approach is the structure where multiple number of gates are used to control the channel carriers. This device architecture is accordingly called multiple-gate MOSFETs (MuGFETs). This structure has the benefit of decreasing the SCEs and increasing the subthreshold slope, as well as providing higher packing densities. The significance of subthreshold performance is tied to the minimizing power requirement in low power portable applications. With improvement of the subthreshold slope, less threshold voltages can be applied to maintain the drain current drive without enhancing leakage currents. Unfortunately, multiple-gate semiconductor devices are disadvantaged by process complexity. Still, as of today, MuGFETs are recognized as having the best potential among the advanced device structures to maintain the node-to-node performance improvement predicted by Intel scientist G. Moore.

The salient features of the MuGFET are [69,70]:

(i) The multi-gate-FETs can control the SCEs by the modified physical structure of device, compared to that of conventional MOSFET in which the SCEs are mitigated by a special type of doping technique (halo doping or channel doping); (ii) a very thin layer of silicon as channel leads to better gate potential coupling with the potential of channel.

These important properties have potential benefits that involves

1. The less SCEs prominent to a small length of channel associated to conventional MOSFET
2. A shriller subthreshold slope (SS) (~60 mV/dec but for conventional MOSFET SS is ~80 mV/dec) thus, permits a higher gate overdrive current for the equivalent applied power supply and the equivalent off-state leakage current;
3. Multi-gate device has good carrier transport because the channel impurities are reduced (in advanced device structures, the doping less channel can be employed).

Channel impurities reduction also dismisses a scaling problems due to band-to-band and the drain-to-body tunneling leakage current. The main benefit is high drain current drive (or total gate capacitance) per semiconductor area of the device. Carrier flow in the multi gate FET with doping less channel is greater than that in bulk MOSFETs for two reasons:

(a) The less Coulomb scattering because of lower ionization of dopants in the doping less channel area, and less surface roughness scattering because of lower electric field at the surface. In conventional MOSFETs, doping of the channel is applied to set the variation in threshold voltage and pocket or

halo doping is applied to lower the SCEs. These ionized depletion charges in the device add to the surface electric field.

In a MuGFET with a doping less or less doped channel, no ionized depletion charge is present, thus, the electric field at the surface is subsidized completely by inversion carriers in the device. Although the mobility of carrier in semiconductor devices follows the "universal mobility" curve, with equal gate overdrive, the mobility of carrier can be expressively higher when compared to a bulk MOSFET because the effective electric field is less at the equal gate overdrive. While the conventional bulk MOSFET works at an effective field higher than 1 mV/cm, the MuGFET with a doping less or less doped channel works nearby 0.5 mV/cm, thus, enlightening the mobility to two times virtually. This improves transfer of charges because in the DG MOSFET, capacitance C is doubled, and the current I is enhanced by above two times due to improved transport. The MuGFET is very favorable for downscaling below 50 nm. The main motives overdue this assumption are given as the following

1. The drain drive current rises in the subthreshold region as $\exp\left[q\left(V_{gs}-V_{th}\right)/\left(K_bT\right)\right]$, with an ideality factor equal to 1, because of the almost ideal subthreshold slope of 60 mV/decade. This is very important to reduce the leakage current at low threshold voltages;
2. The protection caused by extra gates makes MuGFET more resistant to the SCEs as compared to conventional MOSFET or SOI MOSFET;
3. The shielding impact of the extra gates avoids punch through to take place even with at no doping within the channel, which is essential to prevent degradation of the output transfer characteristics and extra leakage currents with gate voltage $V_{gs=}0\,V$;
4. The drain drive current enhanced with velocity exceed is exactly considerable at the dimensional limits granted by the MuGFET. Also, the transit time across the channel length could be kept smaller than 1ps, which enhanced the performance of a MOSFET with a similar gate thickness;
5. For a given semiconductor device size, the MuGFET transconductance is in any case double substantial to that of any industrial standard MOSFET due to the effect of at least two gates.

As per discussion above, the MuGFET is predictable to provide high performance in digital VLSI circuit design with the condition that the resistance of source/drain can be kept less than the intrinsic structure resistance. Due to thin silicon channel, the series resistance of the MuGFET is important to consider. Some type of an increased S/D process would be employed in a manner to attain a S/D fan-out. On thin silicon substrates (or fins) (<15-nm), growing epitaxial film is hard, due to the thin starting silicon film tends to crack during or proceeding to the epitaxial growth. In addition

to this, as the thickness of silicon channel is minimized by approximately 10 nm, optimization of the semiconductor device's parasitic capacitance, and series resistance may demonstrate it difficult.

4.4 FINFET DEVICE STRUCTURES

FinFET architecture (Figure 4.4) originating from a previous device known as a FD lean channel transistor (DELTA) was originally presented in 1989. The concept of multiple-gate FETs was already familiar [71,72], but the novelty of the DELTA implementation resulted in the development of similar devices.

Same principle is behind the function of both DELTA and FinFET transistors. The body of the device (fin) is a relatively thin architecture that linked the large S/D pads. With a gate dielectric created on the side of the fin, a conformal material of gate is placed to cover both sides of the fin, creating a tied DG transistor. The current conduction is thus on the side of the fin that linked S/D, and the width of the device channel is usually approximated as two times the fin height. The device drive current is simply increased by addition of extra parallel fins to the architecture as permissible by the S/D dimensions. Although familiar for over a decade, these devices have garnered a lot of attention as of late. There have been several groups working on FinFET devices in recent years with most research coming from University of California at Berkeley (UCB), IBM, and Intel.

Due to the unconventional nature of the device (sidewall channel conduction), transistor metrics can be interpreted in several ways. It is worthy to note that for FinFETs, channel width is often defined by fin height. However, when compared to classical MOSFETs, the channel width that should be compared is actually twice the fin height. These concepts should be kept in mind when evaluating device performance from the literature. The approaches taken in fin creation employ a number of techniques: from

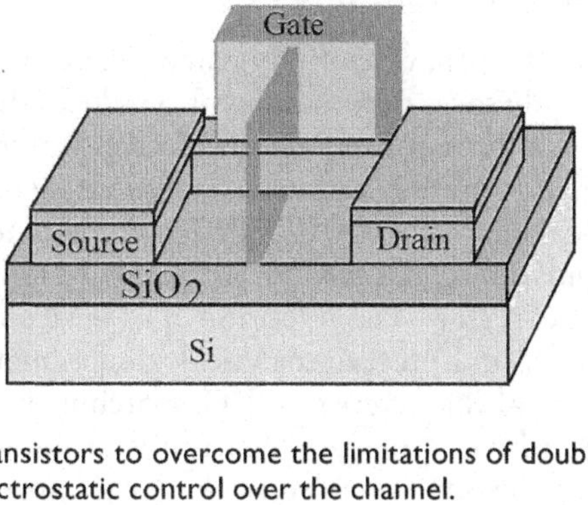

Figure 4.4 FinFET transistors to overcome the limitations of double-gate MOSFET and better electrostatic control over the channel.

electron-beam lithography to conventional deep ultraviolet photolithography and fusion of ultraviolet photolithography and spacer-based lithography. Research has usually concentrated on the growth of ultra-thin body devices to make sure full depletion with various efforts related to reducing etch damage in the channel region.

4.4.1 FinFET's pros and cons

The multiple research efforts in the FinFET have already produced impressive results from both academia and industry. However, FinFETs are not the only solution to the problems of continued scaling. The first and most obvious approach is to continue with traditional planar CMOS technologies until a fundamental barrier, such as the size of the silicon atom, is achieved. The cost associated with transition to entirely new types of devices is immensely prohibitive considering the time and investment needed to establish new design and manufacturing processes. Thus, the transition to new and riskier solutions is a tremendous undertaking with the industry focusing on an approach that works now. Still, most major manufacturers and researchers have investigated alternatives to both planar CMOS and FinFETs.

The "multi-gate" FET device group, to which FinFETs belong, contains several proposed solutions to scaling problems. For example, researchers have already demonstrated functioning of DG planar devices and gate-all-around (GAA) devices. These types of FETs present benefits similar to those of FinFETs: improved SCEs and subthreshold slope with an increased drive current density. In FinFET, even the effective transistor width is controlled by the number of fins present, planar DG devices are considered to be limited in width to less than a micron, while GAA devices often present tremendous design and process difficulties in manufacturing. Issues such as top-and-bottom gate size matching and alignment and parasitic capacitance make process integration difficult.

4.4.2 FinFET device modeling

Precisely, predicting the behavior of fabricated devices using device models and simulators not only saves time and but also reduces the cost. As a result, FinFET modeling is a current topic of advanced research for many academician, industrialist, engineers, and device physicists working in this area. Simple 2-D analytical models of the MOSFET is required for computer-aided design of digital and analog ICs having thousands to millions semiconductor devices on a silicon chip. The importance of device modeling is developing simple, reliable, and precise analytical model (mathematical model) that represents electrical characteristics (DC, switching behavior, and small-signal analysis) of FinFETs. Compact DG transistor models are required to compute analytically the transistor electrical analysis, speedily further use in

circuit simulators to develop and optimize design for high performance of silicon monolithic ICs (or chips) having thousands to millions of same type or different transistors for switch and other analog/digital applications. Earlier design focuses were on high performance only, but in 1990s, researcher's interest in low-power, low-voltage circuit design was developed because of more portable applications and led efforts to provide designers with better transistor mathematical models and design methods using simulation tools. The Enz, Krummenacher and Vittoz (EKV) model developed around 1995, evaluated MOSFET behavior in weak, moderate, and strong inversion, permitted device designers to develop the accuracy of their manual calculations and simulations and analysis of low-voltage, low-power circuit design.

Analytical modeling and device structure design of DG FinFET with various materials have been reported in the literature by many researchers. The DG FinFET operation and inversion charge have been evaluated by Munteanu et al. [73]. Raj et al. [60] presented a generic hidden surface potential analysis for undoped DG FinFET, physically scalable applied gate biases and insulator/channel thickness variations and catering to applied gate biases to both gates using one-dimensional (1-D) Poisson's equation approach based on the classical domain. For nanoscale device dimension, we require 2-D approach for DG FinFET devices which will be valid for classical as well as in quantum domains.

The quantum surface potential can be developed with the consideration of quantum mechanical effects (QMEs) which play a major role within the device for such a nanoscale dimension [73]. SCEs in advanced nanoscale DG FinFETs have been studied widely [16] and show the requirement for UTB thickness less than or approximate to the De Broglie wavelength. Modeling of QMEs in nanoscale devices has required less interest, due to inerrant properties of these devices and having less leakage current. Since, charge carrier in nanoscale devices is subjected to structural confinement (SC) of the device under study; also, device field-induced electrical confinement (EC) [7] and QMEs on the subthreshold electrostatics must be considered. When channel charge carriers are spatially confined in one dimension, by either SC or EC, charge carrier energy quantization for DG FinFETs becomes significant for consideration [12].

Threshold voltage is a vital device parameter that governs switching of any FET device. Various modeling approaches of threshold voltage have been discussed [16] for the DG FinFET device. Katti et al. [33] reported the evaluation of threshold voltage based on pure quantum charge. The model developed by Chiang et al. [74] is a 2-D approach of threshold voltage evaluation with the consideration of least sheet density of inversion charge carriers which reaches a value of threshold charge sufficient for identifying the ON state, and it is a fully classical approach. Raj et al. approach [7] is based on physics-based threshold voltage for SOI MOSFET while Wong et al. [39] simulated on FIELDAY simulation for the evaluation of threshold voltage of the undoped channel DG FinFET.

S/D extension region resistances are an important concern when designing such structures because these resistances can be very high (due to the use of an ultra-thin body), thus limiting device performance. Analysis of the parasitic and total S/D resistance in the FinFET device has been presented by Dixit et al. [3]. The use of doping less ultra-thin bodies also diminishes the problem of adjusting threshold voltage V_{th} by changing the body doping. Gate stack analysis has to be performed to acquire an appropriate V_{th} either by employing various types of terminal materials with attractive work functions, or maintain an off-state voltage between the different gate electrodes to imitate different work functions [32]. QMEs (sub band splitting) play an important role in the confinement of charge carriers becoming sufficient high within the ultra-thin bodies, which leads to an improved sensitivity of V_{th} to the silicon body thickness. This effect proves to be an additional problem of fluctuations in T_{fin} that has to be strictly controlled.

The inversion charge modeling using 1-D potential approach has been discussed analytically as well as numerically [26]. Discussion on inversion charge with varying device parameters to achieve the volume inversion in the channel has been presented by Munteanu et al. [73]. A very few research papers have anticipated analytical inversion of charge carriers using 2-D investigation design approaches. The drain current models have been carried out with QME, including the design approach [4]. In most of mathematical drain current analytical models, it has been reported that the estimation of drain current has been carried out with considering the self-consistent potential of devices within the active silicon region. Accurate gate capacitance modeling has been carried out by Lazaro et al. [4] and electron mobility for DG FinFET by Raj and et al. [7].

To be able to better predict the I-V characteristics of FinFET devices, accurate mobility models are required to incorporate all of the basic scattering mechanism operating in the inversion layer. Numerous models for charge carrier mobility, model for inversion layer, and bulk silicon have been studied in detail in the literature but most of them suffer with some limitation, for example, the validity of the temperature range [34], the difficulty of implementation in simulators of completely generalized (nonplanar) devices. It has also reported that the charge carrier electron and hole motilities in the inversion layer on a (100) surface pursue the universal characteristics at room temperature and also not depend on the substrate doping concentration or the external bias when analyzed as a function of effective normal fields [2]. Esseni et al. [13] reported that the normal field dependence of the electron mobility is well explained in terms of electron quantization.

4.5 BULK MOSFET VS FINFET LEAKAGE CURRENTS

Reducing the thickness of the gate oxide at each technology node can help to improve the on-state current of a MOSFET. Larger 'on state' current

can charge the capacitors of the MOSFET more easily reducing the device delays. As the oxide thickness of gate becomes fantastically thin, the gate oxide will be the constraints of scaling MOSFETs due to high gate leakage current. First, electrons and holes continuously leak through such thin gate oxide. In other words, the leakage problems become serious and power dissipation will increase a lot. The leakage currents in nanoscale devices have been described by number of authors [23]. Choi et al. [33] described the leakage current in FinFET with scaling theory. The possible solution to minimize the leakage current is to use other dielectric materials instead of the traditional SiO_2 dielectric layer with higher dielectric constant layer than silicon dioxide, and the other is an alternative advanced MOS structure like FinFET. The major contribution of leakage current in FinFET devices is due to subthreshold leakage current and gate leakage current.

4.5.1 Subthreshold leakage currents

The dominant components of static power dissipation in VLSI chips are subthreshold leakage currents and power dissipation. It also creates problem in switching the device off. Due to small current, there will be a small voltage present at the output node of CMOS circuits. The major components of subthreshold currents are DIBL and diffusion of electrons from source to drain below the threshold voltage. Chin et al. [75] described the subthreshold leakage current in FinFET and its scaling limits. The subthreshold currents are very sensitive to changes in V_{th} and thus with threshold voltage scaling, subthreshold leakage currents will increase exponentially. The impact of DIBL is to lower the barrier and thus more electrons are able to cross the barrier and run to drain side, thus producing a small leakage current even if gate voltage is lower than that of threshold voltage. The other component comes from diffusion of electrons from source to drain as diffusion current is proportional to $\partial n/\partial y$, where n is electron concentration. With decrease in the feature size (i.e., channel length), the gradient of charge increases resulting in the increase of leakage current. Figure 4.5 gives the variation of subthreshold current with gate voltage. The subthreshold current has been calculated from the subthreshold slope (mV/decade of current), as shown in Figure 4.5.

4.5.2 Gate leakage current

The downscaling of bulk MOSFETs has been required in reduction of the oxide thickness to obtain higher drain current driving capability and to improve control over the SCEs. It has been predicated by ITRS that gate oxide thicknesses (T_{ox}) of 1–1.2 nm will be necessary requirement for sub-50 nm MOSFETs. Due to thin gate oxide under the gate terminal, direct tunneling current increases sharply with decreased oxide thickness (T_{ox}) that is of major concern for the conventional MOSFET downscaling. Gate

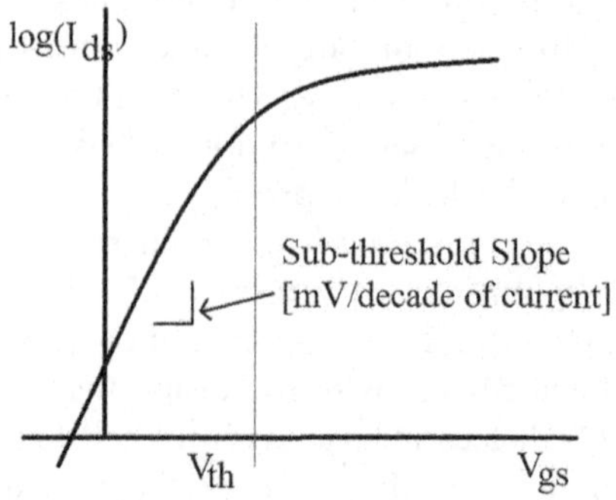

Figure 4.5 Variation of sub-threshold current with gate voltage to analysis the sub-threshold slope (SS) of semiconductor device.

direct tunneling leakage current evolved from the quantum effect of a charged carrier through the thin gate oxide barrier into the gate terminal that relies not only on the MOSFET structure but also depends on the external biasing conditions.

It is obvious that in advanced nanoscale devices, the various types of leakage currents grow to be important, and the level of current depends highly on the device geometry and doping profile. DG FinFET has been talented for downscaling semiconductor devices into the nanometer regime without the increase in leakage current [11]. In the DG FinFET device, because of outstanding control of the SCEs, there is a lower subthreshold leakage current.

4.5.3 Robotic application

As discussed, nanoscale semiconductor have acquired higher attention in robotic systems for their novel electronic, mechanical, photonic, thermal, and electrical properties. Nanoscale semiconductor devices having 1-D structure hold large surface to volume ratio that can be employed in the effective confinement of light energy and thus this exceptional property is desired in the designing of extremely sensitive photodetectors, solar cells, and optical interconnects for robotic systems. As compared to indirect band gap semiconductor (e.g., Si), direct band gap semiconductor nanowire III-V compound such as Gallium Arsenide (GaAs) has benefit of absorbing light energy more efficiently. MOSFETs made of GaAs as the channel material also offers extremely high mobility of carrier. As bare GaAs has large density of surface states because of its inherent geometry, which in turn severely degrades device characteristics by pinning the surface Fermi energy. So in order to prevent device degradation in bare GaAs, it is coated with the shell of $Al_xGa_{1-x}As$ that passivates the non-radiative charge traps which decreases surface scattering results in higher carrier transport efficiency for sensor used in robotic applications.

4.6 SUMMARY

In this chapter, we have carried out literature review of FinFET devices and various modeling techniques required for robotic systems. The reported literature study is useful to recognize various technical research gaps in nanoscale semiconductor devices area of research. Through this study, we have analyzed various devices structures to connect the technical research gaps in order to have a good device in terms of high speed and low power for parable applications in robotic systems. Many research articles, papers, monographic, and theses have also been referred to in the area of advanced nanoscale semiconductor device simulations and sensor design. Articles on implementation of FinFET sensors that is having low leakage, high speed, and high reliability were also studied for robotic systems.

REFERENCES

1. Agarwal, A., Mukhopadhyay, S., Raychowdhury, A., Roy, K. and Kim, C.H. 2006. Leakage power analysis and reduction for nanoscale circuits. *IEeE Micro*, *26*(2), pp. 68–80.
2. Raj, B., Saxena, A.K., and Dasgupta, S. 2009. Analytical modeling for the estimation of leakage current and subthreshold swing factor of nanoscale double gate FinFET device. *Microelectronics International*, 26(1), pp. 53–63.
3. Dixit, A., Kottantharayil, A., Collaert, N., Goodwin, M., Jurczak, M., and De Meyer, K. 2005. Analysis of the parasitic S/D resistance in multiple-gate FETs. *IEEE Transactions on Electron Devices*, *52*(6), pp. 1132–1140.
4. Lázaro, A., Nae, B., Moldovan, O., and Iñiguez, B. 2006. A compact quantum model of nanoscale double-gate metal-oxide-semiconductor field-effect transistor for high frequency and noise simulations. *Journal of Applied Physics*, *100*(8), p. 084320.
5. Ortiz Conde, A., Sánchez, F. J. G., Schmidt, P. E., and Sa Neto, A. 1989. The non-equilibrium inversion layer charge of the thin-film SOI MOSFET. *IEEE Transactions on Electron Devices*, *36*(9), pp. 1651–1656.
6. Doyle, B., Boyanov, B., Datta, S., Doczy, M., Hareland, S., Jin, B., Kavalieros, J., Linton, T., Rios, R., and Chau, R. 2003. Tri-gate fully-depleted CMOS transistors: Fabrication, design and layout. In *2003 Symposium on VLSI Technology. Digest of Technical Papers (IEEE Cat. No. 03CH37407)* (pp. 133–134). IEEE.
7. Raj, B., Saxena, A.K., and Dasgupta, S. 2008. A compact drain current and threshold voltage quantum mechanical analytical modeling for FinFETs. *Journal of Nanoelectronics and Optoelectronics*, *3*(2), p. 163.
8. Cherkauer, B.S. and Friedman, E.G. 1995. A unified design methodology for CMOS tapered buffers. *IEEE Transactions on Very Large Scale Integration (VLSI) Systems*, *3*(1), pp. 99–111.
9. Doyle, B.S., Datta, S., Doczy, M., Hareland, S., Jin, B., Kavalieros, J., Linton, T., Murthy, A., Rios, R., and Chau, R. 2003. High performance fully-depleted tri-gate CMOS transistors. *IEEE Electron Device Letters*, *24*(4), pp. 263–265.

10. Yu, B., Chang, L., Ahmed, S., Wang, H., Bell, S., Yang, C.Y., Tabery, C., Ho, C., Xiang, Q., King, T.J., and Bokor, J. 2002. FinFET scaling to 10nm gate length. In *Digest. International Electron Devices Meeting* (pp. 251–254). IEEE.
11. Raj, B., Mitra, J., Bihani, D.K., Rangharajan, V., Saxena, A.K., and Dasgupta, S. 2011. Process variation tolerant FinFET based robust low power sram cell design at 32nm technology. *Journal of Low Power Electronics*, 7(2), pp. 163–171.
12. Dennard, R.H. 1984. Evolution of the MOSFET dynamic RAM—A personal view. *IEEE Transactions on Electron Devices*, *31*(11), pp. 1549–1555.
13. Esseni, D., Mastrapasqua, M., Celler, G.K., Fiegna, C., Selmi, L. and Sangiorgi, E. 2003. An experimental study of mobility enhancement in ultra-thin SOI transistors operated in double-gate mode. *IEEE Transactions on Electron Devices*, *50*(3), pp. 802–808.
14. Singh, K. and Raj, B. 2015. Temperature-dependent modeling and performance evaluation of multi-walled CNT and single-walled CNT as global interconnects. *Journal of Electronic Materials*, *44*(12), pp. 4825–4835.
15. Hisamoto, D., Lee, W.C., Kedzierski, J., Anderson, E., Takeuchi, H., Asano, K., King, T.J., Bokor, J., and Hu, C. 1998. A folded-channel MOSFET for deep-sub-tenth micron era. *IEDM Tech. Dig*, *38*(1998), pp. 1032–1034.
16. Hisamoto, D., Lee, W.C., Kedzierski, J., Takeuchi, H., Asano, K., Kuo, C., Anderson, E., King, T.J., Bokor, J., and Hu, C. 2000. FinFET-a self-aligned double-gate MOSFET scalable to 20nm. *IEEE Transactions on Electron Devices*, *47*(12), pp. 2320–2325.
17. Frank, D.J. 2002. Power-constrained CMOS scaling limits. *IBM Journal of Research and Development*, *46*(2.3), pp. 235–244.
18. Chang, J., Huang, M., Shoemaker, J., Benoit, J., Chen, S.L., Chen, W., Chiu, S., Ganesan, R., Leong, G., Lukka, V., and Rusu, S. 2007. The 65-nm 16-MB shared on-die L3 cache for the dual-core Intel Xeon processor 7100 series. *IEEE Journal of Solid-State Circuits*, *42*(4), pp. 846–852.
19. Oh, J. E., Bhattacharya, P. K., Singh, J., Dospassos, W., Clarke, R., Mestres, N., Merlin, R., Chang, K., and Gibala, R. 1990. Epitaxial-growth and characterization OfGaAs/Al/GaAsHeterostructures. *Surface Science*, *228*(3), pp. 16–19.
20. Colinge, J.P. 2004. Multiple-gate soimosfets. *Solid-state electronics*, *48*(6), pp.897–905.
21. Huang, X., Lee, W.C., Kuo, C., Hisamoto, D., Chang, L., Kedzierski, J., Anderson, E., Takeuchi, H., Choi, Y.K., Asano, K., and Subramanian, V. 2001. Sub-50nm P-channel FinFET. *IEEE Transactions on Electron Devices*, *48*(5), pp. 880–886.
22. Marathe, V. G., Paily, R., Dasgupta, N., and Dasgupta, A. 2005. A model to study the effect of selective anodic oxidation on ultrathin gate oxides. *IEEE Transactions on Electron Devices*, *52*(1), pp. 118–121.
23. Sharma, V.K., Pattanaik, M., and Raj, B. 2014. ONOFIC approach: Low power high speed nanoscale VLSI circuits design. *International Journal of Electronics*, *101*(1), pp. 61–73.
24. Hu, W., Chen, X., Zhou, X., Quan, Z., and Wei, L. 2006. Quantum-mechanical effects and gate leakage current of nanoscale n-type FinFETs: A 2d simulation study. *Microelectronics Journal*, *37*(7), pp. 613–619.

25. Chang, H. and Sapatnekar, S.S. 2005. Full-chip analysis of leakage power under process variations, including spatial correlations. In *Proceedings of the 42nd Annual Design Automation Conference* (pp. 523–528).
26. Pei, G., Kedzierski, J., Oldiges, P., Ieong, M., and Kan, E.C. 2002. FinFET design considerations based on 3-D simulation and analytical modeling. *IEEE Transactions on Electron Devices*, *49*(8), pp. 1411–1419.
27. Raj, B., Saxena, A.K., and Dasgupta, S. 2013. Quantum mechanical analytical modeling of nanoscale DG FinFET: Evaluation of potential, threshold voltage and source/drain resistance. *Materials Science in Semiconductor Processing*, *16*(4), pp. 1131–1137.
28. Choi, Y.K., Ha, D., King, T.J., and Hu, C. 2001. Nanoscale ultrathin body PMOSFETs with raised selective germanium source/drain. *IEEE Electron Device Letters*, *22*(9), pp. 447–448.
29. Choi, Y.K., Asano, K., Lindert, N., Subramanian, V., King, T.J., Bokor, J., and Hu, C. 1999. Ultra-thin body SOI MOSFET for deep-sub-tenth micron era. In *International Electron Devices Meeting 1999. Technical Digest (Cat. No. 99CH36318)* (pp. 919–921). IEEE.
30. Brews, J.R., Fichtner, W., Nicollian, E.H., and Sze, S.M. 1979. Generalized guide for MOSFET miniaturization. In *1979 International Electron Devices Meeting* (pp. 10–13). IEEE.
31. Vishvakarma, S.K., Raj, B., Saxena, A.K., and Dasgupta, S. 2007. Modeling of the inversion charge density in the nanoscale symmetric double gate MOSFET: An analytical approach. *Journal of Nanoelectronics and Optoelectronics*, *2*(3), pp. 287–293.
32. Ha, D., Ranade, P., Choi, Y.K., Lee, J.S., King, T.J., and Hu, C. 2003. Molybdenum gate work function engineering for ultra-thin-body silicon-on-insulator (UTB SOI) MOSFETs. *Japanese Journal of Applied Physics*, *42*(4S), p. 1979.
33. Katti, G., DasGupta, N., and DasGupta, A. 2004. Threshold voltage model for mesa-isolated small geometry fully depleted SOI MOSFETs based on analytical solution of 3-D Poisson's equation. *IEEE Transactions on Electron Devices*, *51*(7), pp. 1169–1177.
34. Choi, J. H., Bansal, A., Meterelliyoz, M., Murthy, J., and Roy, K. 2006. Leakage power dependent temperature estimation to predict thermal runaway in FinFET circuits. In *IEEE, Proceeding of International Conference Computer Aided Design (ICCAD)*, Association for Computing Machinery New York, United States, (pp. 583–586), November 5–9.
35. Gopal, M., Prasad, D.S.S., and Raj, B. 2013. 8T SRAM cell design for dynamic and leakage power reduction. *International Journal of Computer Applications*, *71*(9), pp. 43–48.
36. Balasubramanium, J.Y.S. 2008. Design of sub-50nm FinFET based low power SRAMs. *Semiconductor Science and Technology*, *23*, p. 13.
37. Frank, D.J., Taur, Y., and Wong, H.S. 1998. Generalized scale length for two-dimensional effects in MOSFETs. *IEEE Electron Device Letters*, *19*(10), pp. 385–387.
38. Kumar, S. and Raj, B. 2015. Compact channel potential analytical modeling of DG-TFET based on Evanescent-mode approach. *Journal of Computational Electronics*, *14*(3), pp. 820–827.

39. Wong, H.S.P., Taur, Y., and Frank, D.J. 1998. Discrete random dopant distribution effects in nanometer-scale MOSFETs. *Microelectronics Reliability, 38*(9), pp. 1447–1456.
40. Singh, G., Sarin, R.K., and Raj, B. 2016. A novel robust exclusive-OR function implementation in QCA nanotechnology with energy dissipation analysis. *Journal of Computational Electronics, 15*(2), pp. 455–465.
41. Zhang, R. and Roy, K. 2002. Low-power high-performance double-gate fully depleted SOI circuit design. *IEEE Transactions on Electron Devices, 49*(5), pp. 852–862.
42. Yan, R.H., Ourmazd, A., and Lee, K.F. 1992. Scaling the Si MOSFET: From bulk to SOI to bulk. *IEEE Transactions on Electron Devices, 39*(7), pp. 1704–1710.
43. Chang, L., Choi, Y.K., Ha, D., Ranade, P., Xiong, S., Bokor, J., Hu, C., and King, T.J. 2003. Extremely scaled silicon nano-CMOS devices. *Proceedings of the IEEE, 91*(11), pp. 1860–1873.
44. Yin, C., Chan, V.W., and Chan, P.C. 2001. Low S/D resistance FDSOI MOSFETs using polysilicon and CMP. In *Proceedings 2001 IEEE Hong Kong Electron Devices Meeting (Cat. No. 01TH8553)* (pp. 89–92). IEEE.
45. Sharma, S., Raj, B., and Khosla, M. 2016. Comparative analysis of MOSFET, CNTFET and NWFET for high performance VLSI circuit design: A review. *Journal of VLSI Design Tools & Technology (JoVDTT), STM Journals,* 6(2), 19–32.
46. Raj, B., Saxena, A.K., and Dasgupta, S. 2009. Analytical modeling for the estimation of leakage current and subthreshold swing factor of nanoscale double gate FinFET device. *Microelectronics International, 26*(1), 53–63.
47. Wakabayashi, H., Yamagami, S., Ikezawa, N., Ogura, A., Narihiro, M., Arai, K.I., Ochiai, Y., Takeuchi, K., Yamamoto, T., and Mogami, T. 2003. Sub-10-nm planar-bulk-CMOS devices using lateral junction control. In *IEEE International Electron Devices Meeting 2003* (pp. 20–27). IEEE.
48. Choi, Y.K., Lindert, N., Xuan, P., Tang, S., Ha, D., Anderson, E., King, T.J., Bokor, J., and Hu, C. 2001. Sub-20nm CMOS FinFET technologies. In *International Electron Devices Meeting. Technical Digest (Cat. No. 01CH37224)* (pp. 19–1). IEEE.
49. Islam, T., Pramanik, C., and Saha, H. 2005. Modeling, simulation and temperature compensation of porous polysilicon capacitive humidity sensor using ANN technique. *Microelectronics Reliability, 45*(3–4), pp. 697–703.
50. Kita, T., Wada, O., Ebe, H., Nakata, Y., and Sugawara, M. 2002. Polarization-independent photoluminescence from columnar InAs/GaAs self-assembled quantum dots. *Japanese Journal of Applied Physics, 41*(10B), p. L1143.
51. Kushwaha, M., Prasad, B., and Chatterjee, A.K. 2016. Gate tunnelling current model for nanoscale MOSFETs with varying surface potential. *IETE Journal of Research, 62*(3), pp. 347–355.
52. Shao, X. and Yu, Z., 2005. NanoscaleFinFET simulation: A quasi-3D quantum mechanical model using NEGF. *Solid-Stae Electronics, 49*(8), pp. 1435–1445.
53. Polishchuk, I., Ranade, P., King, T.J., and Hu, C. 2004. *Dual work function CMOS gate technology based on metal interdiffusion.* U.S. Patent 6,794,234.

54. Semenov, O., Vassighi, A., and Sachdev, M. 2002. Impact of technology scaling on thermal behavior of leakage current in sub-quarter micron MOSFETs: Perspective of low temperature current testing. *Microelectronics Journal*, *33*(11), pp. 985–994.
55. Jiang, W., Tiwari, V., de la Iglesia, E., and Sinha, A. 2002. Topological analysis for leakage prediction of digital circuits. In *Proceedings of ASP-DAC/VLSI Design 2002. 7th Asia and South Pacific Design Automation Conference and 15h International Conference on VLSI Design* (pp. 39–44). IEEE.
56. Sharma, V.K., Pattanaik, M., and Raj, B. 2015. INDEP approach for leakage reduction in nanoscale CMOS circuits. *International Journal of Electronics*, *102*(2), pp. 200–215.
57. Colinge, J. P. *Silicon-on-Insulator Technology: Materials to VLSI*, Kluwer Academic Publishers, Dordrecht, Netherlands, 1991.
58. Uyemura, J.P. 2002. Introduction to VLSI circuits and systems.
59. Wong, H.S., Chan, K.K., and Taur, Y. 1997. Self-aligned (top and bottom) double-gate MOSFET with a 25 nm thick silicon channel. In *International Electron Devices Meeting. IEDM Technical Digest* (pp. 427–430). IEEE.
60. Jain, N. and Raj, B. 2016. An analog and digital design perspective comprehensive approach on Fin-FET (fin-field effect transistor) technology—a review. *Reviews in Advanced Sciences and Engineering*, *5*(2), pp. 123–137.
61. Sharma, S.K., Raj, B., and Khosla, M. 2016. A Gaussian approach for analytical subthreshold current model of cylindrical nanowire FET with quantum mechanical effects. *Microelectronics Journal*, *53*, pp. 65–72.
62. Chang, L., Tang, S., King, T.J., Bokor, J., and Hu, C. 2000. Gate length scaling and threshold voltage control of double-gate MOSFETs. In *International Electron Devices Meeting 2000. Technical Digest. IEDM (Cat. No. 00CH37138)* (pp. 719–722). IEEE.
63. Chang, L., Choi, Y.K., Kedzierski, J., Lindert, N., Xuan, P., Bokor, J., Hu, C., and King, T.J. 2003. Moore's law lives on [CMOS transistors]. *IEEE Circuits and Devices Magazine*, *19*(1), pp. 35–42.
64. Sharma, V.K., Pattanaik, M., and Raj, B. 2014. PVT variations aware low leakage INDEP approach for nanoscale CMOS circuits. *Microelectronics Reliability*, *54*(1), pp. 90–99.
65. Schaller, R.R. 2004. *Technological innovation in the semiconductor industry: A case study of the International Technology Roadmap for Semiconductors (ITRS)* (Doctoral dissertation, George Mason University).
66. Yuan, J., Zeitzoff, P. M., and Woo, J. C. S. 2002. "Source/drain parasitic resistance role and electrical coupling effect in sub 50 nm MOSFET design," *32nd European Solid State Device Research Conference*, pp. 503–506.
67. Cristoloveanu, S. and Ioannou, D.E. 1990. Adjustable confinement of the electron gas in dual-gate silicon-on-insulator MOSFET's. *Superlattices and Microstructures*, *8*(1), pp. 131–135.
68. Kumar, S. and Raj, B. 2016. Analysis of I ON and Ambipolar current for dual-material gate-drain overlapped DG-TFET. *Journal of Nanoelectronics and Optoelectronics*, *11*(3), pp. 323–333.
69. Singh, A., Khosla, M., and Raj, B. 2016. Comparative analysis of carbon nanotube field effect transistor and nanowire transistor for low power circuit design. *Journal of Nanoelectronics and Optoelectronics*, *11*(3), pp. 388–393.

70. Singh, J. and Raj, B. 2019. Design and investigation of 7T2M-NVSRAM with enhanced stability and temperature impact on store/restore energy. *IEEE Transactions on Very Large Scale Integration (VLSI) Systems*, *27*(6), pp. 1322–1328.
71. Singh, J. and Raj, B. 2019. An accurate and generic window function for nonlinear memristor models. *Journal of Computational Electronics*, *18*(2), pp. 640–647.
72. Bhushan, S., Khandelwal, S., and Raj, B. Analyzing different mode FinFET based memory cell at different power supply for leakage reduction. In *Proceedings of Seventh International Conference on Bio-Inspired Computing: Theories and Applications (BIC-TA 2012)* 2013 (pp. 89–100). Springer.
73. Munteanu, D., Autran, J.L., Loussier, X., Harrison, S., Cerutti, R., and Skotnicki, T. 2006. Quantum short-channel compact modelling of drain-current in double-gate MOSFET. *Solid-State Electronics*, *50*(4), pp. 680–686.
74. Chiang, M.H., Lin, C.N., and Lin, G.S. 2006. Threshold voltage sensitivity to doping density in extremely scaled MOSFETs. *Semiconductor Science and Technology*, *21*(2), p. 190.
75. Chen, Q. and Meindl, J.D. 2004. Nanoscale metal–oxide–semiconductor field-effect transistors: Scaling limits and opportunities. *Nanotechnology*, *15*(10), p. S549.

Chapter 5

Internet of things for smart gardening and securing home from fire accidents

Koppala Guravaiah and R. Leela Velusamy
National Institute of Technology Tiruchirappalli

CONTENTS

5.1 INTRODUCTION

In recent years, smart home design using internet is increasing due to the availability of high speed communication and to add comfort, convenience, and safety for human beings (Lee and Lee 2015; Gil et al. 2016). Internet of things (IoT) plays a major role in the design of smart homes. Wireless sensor network (WSN) is an emerging technology to assist IoT applications.

Recently, there are several applications (Miorandi et al. 2012; Ferdoush and Li 2014), such as smart homes, smart city, smart grids, vehicle monitoring, e-health, smart retail, and smart agriculture, designed using the concept in WSNs (Yick et al. 2008; Akyildiz et al. 2002).

In this chapter, proposal for automating two activities that can convert a home to smart home is discussed. The first activity is to automate the process of protecting home from fire accidents. The second activity is to automate the process of watering the plants in a garden. Fire accident is an unexpected event that could fetch a big loss to social asserts and human life. Fire accidents may occur for different reasons such as short circuit and gas leakage. Fire accidents are reported both in industrial and non-industrial premises such as residential buildings, educational institutions, hotels, commercial complexes, hospitals, and assembly halls.

The second activity is related to gardening. Gardening is one of the best recreation activity for human being. In recent days, due to their busy schedule, people find it difficult to maintain gardens even though they are interested to have one. The main problem faced in maintenance of a garden is watering the plants on time. An automated plant watering system will help them to have a garden of their own.

Data collection is important operation in IoTs and WSNs. In this proposed work, river formation dynamics (RFD)-based data collection (Guravaiah and Velusamy 2015) is used to collect the data from sensors and store them in database for further analysis.

Recent survey (Imteaj et al. 2017; Vijayalakshmi and Muruganand 2017; Pol et al. 2016; Vaishali et al. 2017; Shekhar et al. 2017; Kissoon et al. 2017; Saraf and Gawali 2017) shows that attempts have been made to reduce fire accidents and watering the gardens using WSNs and IoTs. In this this chapter, we discuss the design and development of the following systems:

1. Fire alarm system that can prevent fire accidents
2. Smart garden with automated plant watering system
3. Using RFDMRP: River Formation Dynamics-based Multi-hop Routing Protocol for data collection from sensors to server for further processing.

In this book chapter, we will deliberate upon the above methods used to protect the home from fire accidents and smart garden with a water planting system. The technology/technical terms used in the book chapter are explained in Sections 5.1 and 5.2. Section 5.3 deals with the existing approaches and limitations of those approaches. Section 5.4 explains the proposed system. Section 5.5 concludes the chapter and presents future research directions. Apart from regular references, additional references are also included in the "References for Advance/Further reading" for the benefit of advanced readers.

5.2 INTRODUCTION OF DIFFERENT SENSORS USED

This section will give the introduction to the different sensor components used in the proposed system.

5.2.1 Temperature sensor

The temperature sensor (LM35 Temperature Sensor 2018) used to capture temperature from specific sources and modify the collected information into understandable format for the user is shown in Figure 5.1. Temperature sensors are used in several applications, namely, chemical handling, medical devices, food processing units, environmental controls, controlling systems, etc.

5.2.2 Flame sensor

The flame sensor (Flame 2018) mainly used to detect fire or flame within some geographical area is shown in Figure 5.2. This sensor will be used for fire detection in several applications such as industrial, home, and commercial applications.

5.2.3 Soil moisture sensor

Soil moisture sensor (2018) shown in Figure 5.3 will measure the volumetric water content in soil or land. The soil moisture is sensed using copper electrodes. Measurement of moisture content level is calculated using conductivity between the electrodes. This sensor will be useful for measuring

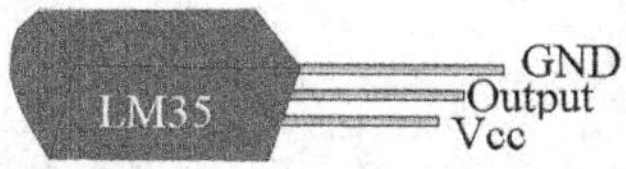

Figure 5.1 Temperature sensor.

Figure 5.2 Flame sensor.

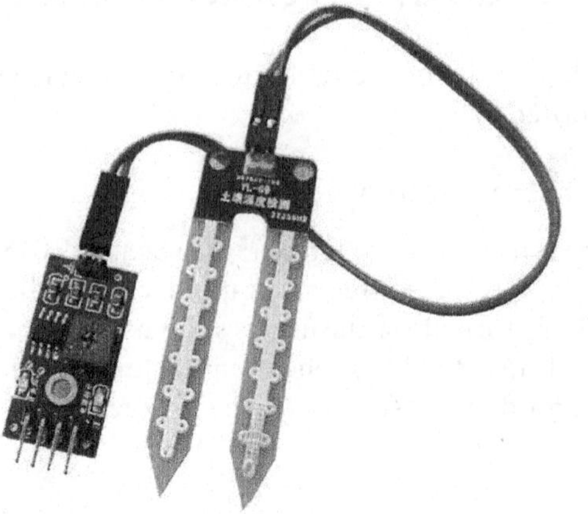

Figure 5.3 Soil moisture sensor.

Figure 5.4 GSM device.

soil moisture in different applications such as soil science, horticulture, botany, environmental science, agricultural science, and biology.

5.2.4 Global system for mobile communication

Global system for mobile communications (GSM) (2018) shown in Figure 5.4 is an international standard for mobile telephones. It is also called as second-generation cellular network (2G). GSM can have outgoing

Figure 5.5 Arduino uno.

and incoming voice calls, simple message system (SMS or text messaging), and data communication (via GPRS).

5.2.5 Arduino uno micro-controller

In a flexible micro-controller development platform called Arduino Uno (2018), as shown in Figure 5.5, ATmega328 micro-controller is used to have easy access of hardware and software components. This micro-controller works with the input voltage of 7–12V operating with 16MHz clock speed and 32K flash memory. In addition, this micro-controller has 14 digital I/O pins and 6 analog inputs.

5.3 SOFTWARE USED

The software used for the following proposed work is RFDMRP, which is explained in the following section.

5.3.1 RFDMRP: River Formation Dynamics-based Multi-hop Routing Protocol

RFD (Rabanal et al. 2007; Amin et al. 2014) is one of the heuristic optimization methods and a subset topic of swarm intelligence, based on replicating the concept of how water drops combine to form rivers and rivers in turn combine to join the Sea by selecting the shortest path based on altitudes of

the land through which they flow. Based on the similarity between RFD and data collection WSN, Guravaiah and Velusamy (2015) proposed a new routing protocol named as RFDMRP for collecting the data from environment and forwarding to base station (BS). This RFDMRP is used for collecting the information from sensors and stored in remote location for further processing.

5.4 EXISTING METHODS

During the last decade, researchers have extensively investigated different applications for designing and managing of smart homes using various micro-processor and micro-controller boards such as Raspberry Pi, Arduino with different sensors such as passive infrared (PIR), camera, temperature, and moisture.

Fire detection system is very important in security of home automation. Not only in home security, but it is also very important in the security of forests and industries. So many methods are implemented for the fire detection system based on image processing and ZigBee technologies. Several alarm systems have been proposed such as smoke detectors and temperature sensor-based systems, and several automated fire alert systems are now available. Some of the relevant proposed works by different researchers are presented in this section.

A Fire Alarm and Authentication System for Workhouse using Raspberry Pi3 is proposed by Imteaj et al. (2017). Garment warehouses and factories are the main focused areas for using the proposed approach. In this system, the camera will take the snapshot of the environment based on the light intensity and gas sensor values, and these will be intimated to the admin by an alarm and SMS.

In the fire WSN (Vijayalakshmi and Muruganand 2017), a node will read the building parameters, namely, fire and temperature, and process the collected data to find out any abnormal conditions. The system takes the appropriate steps based on the collected data.

A methodology based on image processing and sensors is presented by Pol et al. (2016). In this proposed system, a camera is used for detecting images of fire. Initially, the system is trained for recognizing fire, in terms of shape, size, color, intensity, etc., from the automatically captured images. Together with the values from temperature and smoke sensor, the system detects fire, then turns on pump, and also sends the message to local fire control department based on the level of intensity of fire.

Smart irrigation management and monitoring system integrated with a mobile using IoT (Vaishali et al. 2017) is proposed to control the water supply and monitoring the gardens using a smart phone. Intelligent IoT-based automated irrigation system is proposed in the year 2017 by Shekhar et al. (2017). In this work, soil moisture and temperature data are captured, and

accordingly K-Nearest Neighbor classification machine learning algorithm is used for analyzing the sensor data for prediction towards irrigating the soil with water.

In the year 2017, Kissoon et al. (2017) proposed a smart irrigation and monitoring system. In this work, sensors collect air humidity, air temperature, and soil moisture data. These data will be used for monitoring air quality and water content in the soil.

Smart farm irrigation system is proposed by Saraf and Gawali (2017). This proposed system performs remote monitoring and controlling of drips through the WSN using an android phone. In addition to this, sensor nodes and base station are communicated with the help of ZigBee.

In home monitoring, until today the researchers concentrated on home monitoring with different sensors with computational complexities (Suresh et al. 2016; Prathibha et al. 2017). Difficulties present in the literature are as follows:

- Home gardening systems using IoT devices are not presented.
- Fire accident recovery systems are not presented in previous systems.
- Efficient data collection routing protocols are required for efficient data collection.

This chapter concentrates on a simple home monitoring using temperature, flame sensor, and soil moisture sensor with less cost.

5.5 PROPOSED METHODS

This section discusses the effective use of IoT to address the issue of water usage for gardening and monitor gas leakage problems at home or other commercial areas.

5.5.1 Proposed system architecture

A WSN system is a combination of various hardware components and software modules. Figure 5.6 shows the proposed system architecture for home monitoring. The proposed system contains a micro-controller (μC) connected with different sensors such as a gas or flame sensor, a soil moisture sensor, a DC motor, and a GSM module. In the proposed system, as shown in Figure 5.7, different sensors connected with micro-controller will perform the following operations:

- Flame sensing: Any fire generated due to short circuit or gas leakage in house can be sensed by the flame sensor and automatically activate micro-controller to switch on motor to control the fire and send SMS to the corresponding owner of the house.

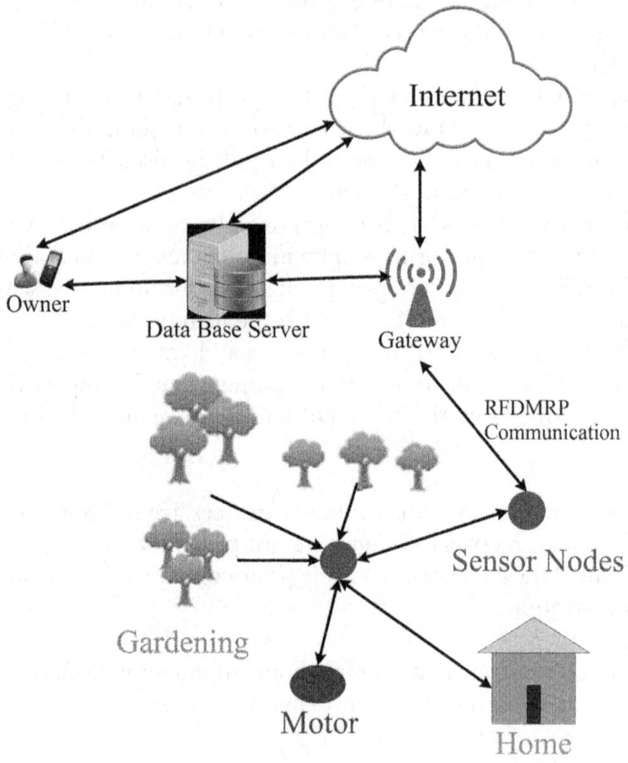

Figure 5.6 Basic model of the proposed approach.

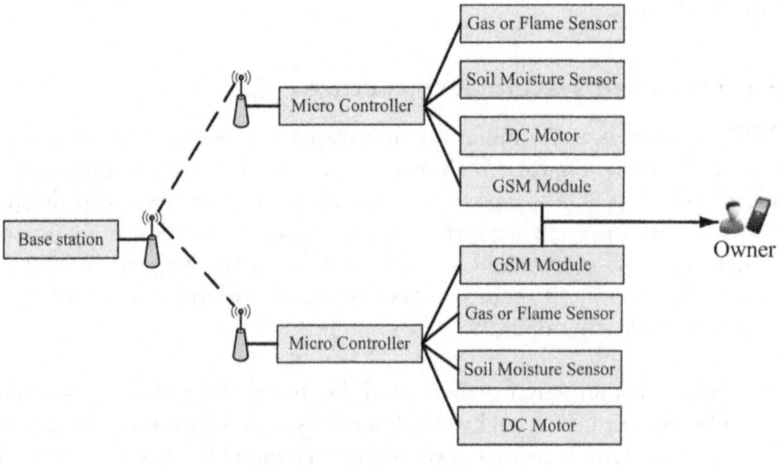

Figure 5.7 Schematic diagram of the proposed system.

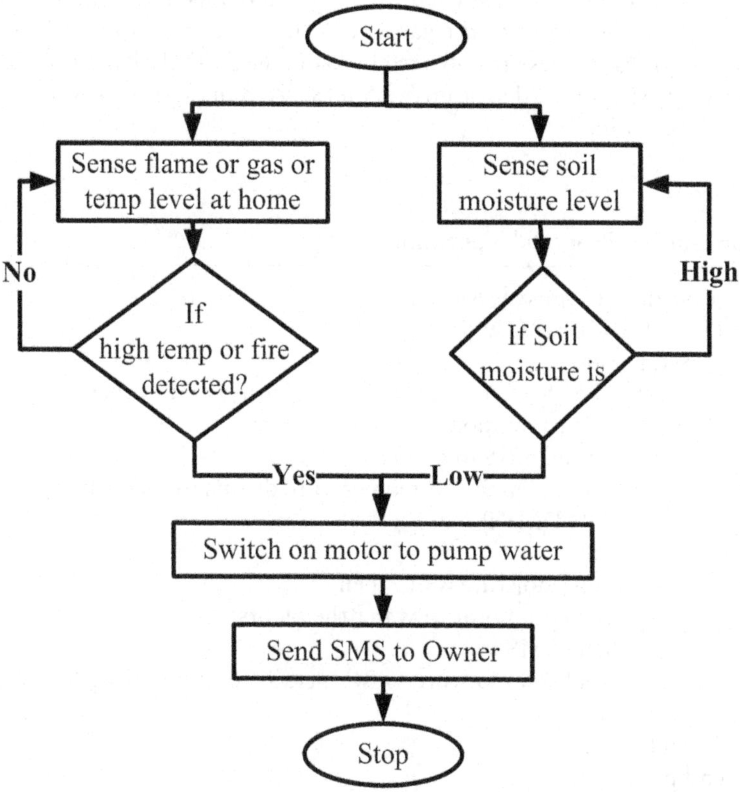

Figure 5.8 Process flow diagram of the proposed approach.

- Temperature sensing: If fire is there, automatically temperature at a particular place will increase. If the temperature is increasing, the temperature sensor will send the details to micro-controller which will activate DC motor and then intimate to the owner of the house about the details.
- Water supply to garden: Soil moisture sensor will sense the soil of the garden. If water level is less, then soil moisture sensor will send the details to micro-controller to activate the GSM to send the message to the owner and start the DC motor to water the plants in the garden.

The process flow diagram of the proposed approach is shown in Figure 5.8.

5.5.2 Proposed algorithm

Whenever flame sensor or temperature sensor values are high, an SMS will be sent to the owner of the house as well as motor is turned on to pump water in order to control the fire. In the same way, when soil moisture

value is low, the motor is started to water the plants and send SMS with the details to the owner of the garden, as shown in Algorithm 5.1. At the same time, these values are collected by using the RFDMRP data collection protocol (as shown in Algorithm 5.2) and send to the server for storing and further processing.

Algorithm 5.1 Proposed algorithm

```
procedure Proposed_Algorithm()
H = HIGH; L = LOW:
    while 1 do
      if ((flame_ Sensor == H) or (Temperature == H)) then
            Turn on motor to pump water
            Send SMS to owner
            Send flame and temperature details to server using
            RFDMRP
        end if
        if (Soil_moisture == L) then
           Turn on motor to water the plants
           Send SMS to owner
           Send soil moisture sensor details to server using RFDMRP
       end if
    end while
end procedure
```

Algorithm 5.2 RFDMRP algorithm

```
procedure RFDMRP()
      nodeDeployment()
      Routers_NNTableCreation()
      EndDevice_NNTableCreation()
      while 1 do
            Sensors read the data from environment
         End devices unicast the packets to router or local server on
detecting sensor values
            repeat
                      Router will unicast the sensor values using
RFDMRP-based multi-hop routing protocol to neighbor routers or
server
            until data reach server
      end while
end procedure
```

5.5.3 Experimental setup and results

For testing purpose, the proposed system is deployed with the help of PIR sensors, temperature sensor, flame sensor, soil moisture sensor, DC motor, and one GSM devise connected to micro-controller. The setup is shown in Figure 5.9.

The messages are received from the micro-controller to the registered mobile phone based on the sensor values and threshold values of the sensors during the sensing time, as shown in Figure 5.10.

The cost of the experimental setup is shown in Table 5.1.

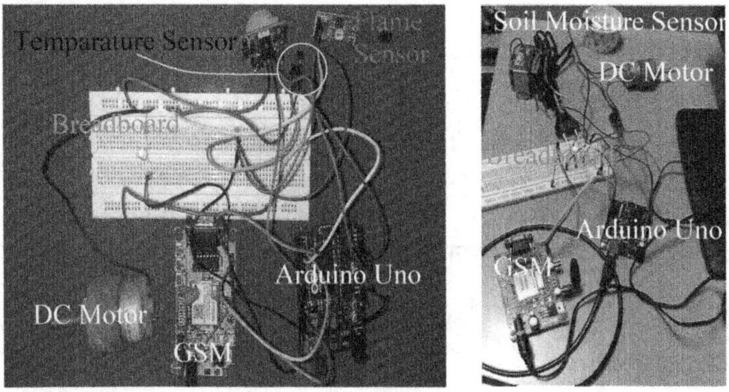

Figure 5.9 Circuit connections of the proposed model.

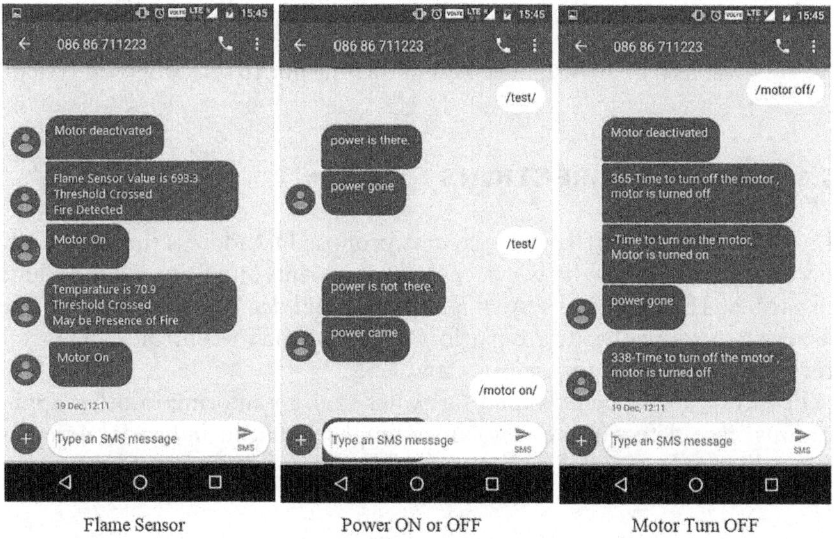

Figure 5.10 Different messages transferred between mobile and sensors.

Table 5.1 Cost of proposed work without additional server

Sl. No.	*Items*	*Cost (in rupees)*
1.	Micro-controller	1,400
2.	GSM module	1,500
3.	Flame sensor	100
4.	Temperature sensor	100
5.	Soil moisture sensor	120
6.	DC motor sensor	1,000
7.	Other expenditure for setup	1,780
	Total	6,000

5.5.4 Applications

The proposed system has the following applications:

- Monitoring of garden or agriculture land for soil water content.
- Monitoring gas and temperature levels at home, laboratories, etc.

5.5.5 Advantages

The proposed system has the following advantages:

- The proposed system is cost effective.
- The water management system is controlled easily using the proposed system.
- Power usage can be reduced for gardening due to this proposed system.

5.6 RESEARCH DIRECTIONS

In this chapter, an IoT-based system is proposed to address the commonly occurring problems in houses such as fire accidents and watering the plants in garden. The proposed system is deployed and demonstrated in real time using open-source hardware platforms such as micro-controller, temperature, soil moisture, flame sensors, and GSM.

In future, it can be extended for water tank monitoring in other applications. Fire detection system can be implemented in industrial area for controlling loss.

REFERENCES

Akyildiz, I. F., W. Su, Y. Sankarasubramaniam and E. Cayirci. 2002. Wireless sensor networks---a survey. *Computer Networks*, 38(4), 393–422.

Amin, S. H., H. Al-Raweshidy, and R. S. Abbas. 2014. Smart data packet ad hoc routing protocol. *Computer Networks*, 62, 162–181.

Ardiuno. 2018. Ardiuno UNO, available at https://www.arduino.cc/en/Main/ArduinoBoardUno/ (Online; accessed 26 December 2018).

Ferdoush, S. and X. Li. 2014. Wireless sensor network system design using Raspberry Pi and Arduino for environmental monitoring applications. *Procedia Computer Science*, 34, 103–110.

Flame. 2018. Flame sensors, available at https://robokits.co.in/sensors/temperature/ame-sensor-module/ (Online; accessed 26 December 2018).

Gil, D., A. Ferrández, H. Mora-Mora, and J. Peral. 2016. Internet of things—a review of surveys based on context aware intelligent services. *Sensors*, 16(7), 1069.

GSM. 2018. GSM device, available at https://www.arduino.cc/en/Guide/ArduinoGSMShield (Online; accessed 26 December 2018).

Guravaiah, K. and R. L. Velusamy. 2015. RFDMRP—river formation dynamics based multi-hop routing protocol for data collection in wireless sensor networks. *Procedia Computer Science*, 54, 31–36.

Imteaj, A., T. Rahman, M. K. Hossain, M. S. Alam, and S. A. Rahat. 2017. An IoT based fire alarming and authentication system for workhouse using Raspberry Pi 3. In *International Conference on Electrical, Computer and Communication Engineering (ECCE)*, 899–904.

Kissoon, D., H. Deerpaul, and A. Mungur. 2017. A smart irrigation and monitoring system. *International Journal of Computer Applications*, 163(8), 39–45.

Lee, I. and K. Lee. 2015. The internet of things (IoT)—applications, investments, and challenges for enterprises. *Business Horizons*, 58(4), 431–440.

LM35 Temperature Sensor. 2018. Temperature sensor, available at http://www.ti.com/lit/ds/symlink/lm35.pdf (Online; accessed 26 December 2018).

Miorandi, D., S. Sicari, F. De Pellegrini, and I. Chlamtac. 2012. Internet of things—vision, applications, and research challenges. *Ad Hoc Networks*, 10(7), 1497–1516.

Pol, S. C., A. H. Wagh, P. T. Ramole, and S. H. Sharma. 2016. Fire detection using image processing and sensors. *International Journal of Engineering Trends and Applications*, 3(2), 85–87.

Prathibha, S. R., A. Hongal, and M. P. Jyothi. 2017. IOT based monitoring system in smart agriculture. In *2017 International Conference on Recent Advances in Electronics and Communication Technology*, 81–84. https://doi.org/10.1109/ICRAECT.2017.52.

Rabanal, P., I. Rodríguez, and F. Rubio. 2007. Using river formation dynamics to design heuristic algorithms. In *Unconventional Computation*, Springer, 163–177.

Saraf, S. B. and D. H. Gawali. 2017. IoT based smart irrigation monitoring and controlling system. In *2nd IEEE International Conference on Recent Trends in Electronics, Information & Communication Technology*, 815–819.

Shekhar, Y., E. Dagur, S. Mishra, and S. Sankaranarayanan. 2017. Intelligent IoT based automated irrigation system. *International Journal of Applied Engineering Research*, 12(18), 7306–7320.

Soil Moisture Sensor. 2018. Soil moisture sensor, available at, https://en.wikipedia.org/wiki/Soil moisture-sensor/. (Online; accessed 26 December 2018).

Suresh, S., J. Bhavya, S. Sakshi, K. Varun, and G. Debarshi. 2016. Home monitoring and security system. In *International Conference on ICT in Business Industry Government (ICTBIG)*, 1–5. https://doi.org/10.1109/ICTBIG.2016.7892665

Vaishali, S., S. Suraj, G. Vignesh, S. Dhivya, and S. Udhayakumar 2017. Mobile integrated smart irrigation management and monitoring system using IOT. In *IEEE International Conference on Communication and Signal Processing (ICCSP)*, 2164–2167.

Vijayalakshmi, S. and S. Muruganand. 2017. Internet of things technology for fire monitoring system. *International Research Journal of Engineering and Technology (IRJET)*, 4(6), 2140–2147.

Yick, J., B. Mukherjee, and D. Ghosal. 2008. Wireless sensor network survey. *Computer Networks*, 52(12), 2292–2330.

REFERENCES FOR ADVANCE/FURTHER READING

Gill, K., S.-H. Yang, F. Yao, and X. Lu. 2009. A ZigBee-based home automation system. *IEEE Transactions on Consumer Electronics*, 55(2), 422–430.

Guravaiah, K. and Velusamy, R.L. 2019. Prototype of home monitoring device using internet of things and river formation dynamics-based multi-hop routing protocol (RFDHM). *IEEE Transactions on Consumer Electronics*, 65(3), 329–338.

Han, D.-M. and J.-H. Lim. 2010. Smart home energy management system using IEEE 802.15.4 and ZigBee. *IEEE Transactions on Consumer Electronics*, 56(3), 1403–1410.

Miorandi. D., S. Sicari, F. De Pellegrini, and I. Chlamtac. 2012. Internet of things—vision, applications and research challenges. *Ad Hoc Networks*, 10(7), 1497–1516.

Chapter 6

Deep CNN-based early detection and grading of diabetic retinopathy using retinal fundus images

Sheikh Muhammad Saiful Islam
Manarat International University

Md Mahedi Hasan
Institute of Information and Communication Technology,
Bangladesh University of Engineering and Technology

Sohaib Abdullah
Manarat International University

CONTENTS

6.1 INTRODUCTION

Diabetic retinopathy (DR), a chronic progressive eye disease, has turned out to be one of the most common causes of vision impairment and blindness especially in people of working ages in the world today [6]. Blood vessels in the light-sensitive tissue (i.e., retina) are mainly affected in DR.

The progress of DR can be categorized into four stages: mild, moderate, severe nonproliferative diabetic retinopathy (NPDR), and the advanced

stage of proliferative DR. In mild NPDR, small areas in the blood vessels of the retina, called microaneurysms (MAs), swell like a balloon. In moderate NPDR, multiple MAs, hemorrhages, and venous beading occur, causing the patients to lose their ability to transport blood to the retina. The third stage, called severe NPDR, results from the presence of new blood vessels, which is caused by the secretion of growth factor. The worst stage of DR is the proliferative DR, as illustrated in Figure 6.1, in which fragile new blood vessels and scar tissue form on the surface of the retina, increasing the likelihood of blood leaking, leading to permanent vision loss.

At present, detection of DR is accomplished through manually detecting vascular abnormalities and structural changes of retina in the retinal fundus images by involving a well-trained physician. These fundus images are taken by dilating the retina using vasodilating agent. Due to the manual nature of DR screening methods, however, highly inconsistent results are found from different doctors, so automated DR diagnosis techniques are essential for solving these problems.

DR can damage retina without showing any indication at the preliminary stage [18]. So, successful early-stage detection of DR can minimize the risk of progression to more advanced stages of DR. The diagnosis is particularly difficult for early-stage detection because the process relies on discerning the presence of MAs, retinal hemorrhages, among other features on the retinal fundus images. Overall, accurate detection and determination of the stages of DR can greatly improve the intervention, which ultimately reduces the risk of permanent vision loss.

Earlier solutions of the automated DR detection system were based on hand-crafted feature extraction and standard machine learning algorithms

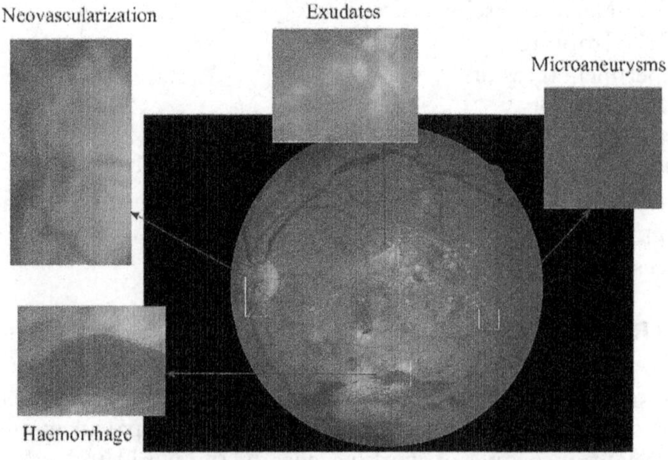

Figure 6.1 Example eye image of the proliferative diabetic retinopathy. Additional new blood vessels will begin to grow on the surface of the retina. Due to their abnormal and fragile nature, retinal hemorrhages and ruptured blood vessels are created in this stage which will lead to permanent vision loss.

for prediction [21]. These approaches greatly suffered due to the hand-crafted nature of DR features extraction since the feature extraction in color fundus images is more challenging compared to the traditional images for the object detection task. Moreover, these hand-crafted features are highly sensitive to the quality of the fundus images, focus angle, presence of artifacts, and noise. Thus, these limitations in traditional hand-crafted features make it important to develop an effective feature extraction algorithm to accurately analyze the subtle features related to the DR detection task.

In recent times, most of the problems of computer vision have been solved with greater accuracy with the help of modern deep learning algorithms, convolutional neural networks (CNNs) being an example. CNNs have been proven to be revolutionary in different fields of computer vision such as object detection and tracking, image and medical disease classification and localization, pedestrian detection, and action recognition. The key attribute of the CNN is that it extracts features in a task-dependent and automated way. So, in this chapter, we present an efficient CNN architecture for DR detection in a large-scale database. Our proposed network has been designed with a multi-layer CNN architecture followed by two fully connected layers and an output layer. Our network outperforms other state-of-the-art networks in early-stage detection and achieves state-of-the-art performance in severity grading of DR.

The rest of the chapter is organized as follows: related work is presented in Section 6.2, followed by the proposed method in Section 6.3, while experimental set-up and results are discussed in Section 6.4. Finally, we draw our conclusion in Section 6.5.

6.2 RELATED WORKS

The earlier works on automatic DR detection were based on designing hand-crafted feature detectors to measure the blood vessels and optic disc, and on counting the presence of abnormalities such as MAs, red lesions, hemorrhages, and hard exudates. The detection was performed using these extracted features by employing various machine learning methods like support vector machines (SVM) and k-nearest neighbor (kNN) [21,23]. In Ref. [1], Acharya et al. used features of blood vessel area, MAs, exudes, and hemorrhages with an SVM, achieving an accuracy of 86%, specificity of 86%, and sensitivity of 82%. Roychowdhury et al. [20] developed a two-step hierarchical classification approach, where the non-lesions or false positives were rejected in the first step. For lesion classification in the second step, they used classifiers such as the Gaussian mixture model, kNN, and SVM. They achieved a sensitivity of 100%, specificity of 53.16%, and AUC of 0.904. However, these types of approaches have the disadvantage of utilizing a limited number of features.

Deep learning-based algorithms have become popular in the last few years. For example, standard ImageNet architectures were used in Refs. [19,27]. Furthermore, Kaggle [11] has recently launched a DR detection competition, where all the top-ranked solutions were implemented employing CNN as the key algorithm. Pratta et al. [19] developed a CNN-based model, which surpassed human experts in classifying advanced stages of DR. In Ref. [2], CNN-based method was employed to detect MAs in a DR stage grading. Ensemble of CNN was employed to simultaneously detect DR and macular edema by Kori et al. [15]. They employed a variant of ResNet [8] and densely connected networks [9]. To make the model prediction more interpretable, a visual map was generated by Torre et al. [25] using a CNN model, which can be used to detect lesion in the tested retinal fundus images. A similar approach was used in Ref. [26] along with the generation of regression activation map (RAM).

Some research studies focused on exploring breakdown of the classification task into subproblem prediction tasks. For example, Yang et al. [28] employed a two-stage deep CNN-based methodology, where exudates, MAs, and hemorrhage were first detected by the local network, and subsequent severity grading was performed by the global network. By introducing unbalanced weight map to emphasize leison detection, they achieved AUC of 0.9590. The authors of Ref. [4] implemented an architecture like VGG-16 [22] and Inception-4 [24] network for DR classification.

Some recent works [5] in DR have leveraged mean squared error objective function to convert the classification task into a regression task. Here, ensemble of classical machine learning algorithms, like naive bayes classifier, SVM, as well as ImageNet state-of-the-art networks, with mean squared error objective function applied to tackle DR detection problem with accuracy, Kappa, and F-score of 0.736, 0.676, and 0.417, respectively.

6.3 PROPOSED METHOD

6.3.1 Data preprocessing

There exists a lot of exposure and lighting variation over the original fundus images, so we took several preprocessing steps, as suggested by Graham [10], to standardize the image condition. First, we re-scaled the images to get the same radius and subtracted the local average color. Thereafter, the local average color of the images was mapped to 50% gray. We also clipped the images to 90% of original size to remove the boundary effects. The sample of the resulted preprocessed images, along with the original images, is illustrated in Figure 6.2.

6.3.2 Data augmentation

The performance of a deep neural network is strongly correlated with the size of available training data. Although the Kaggle EyePACS is the largest

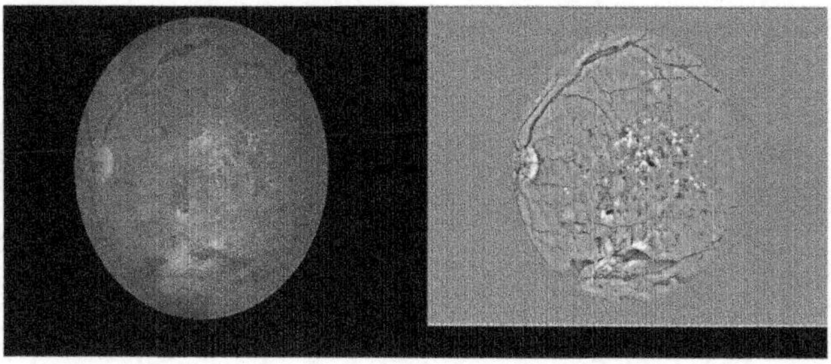

Figure 6.2 Demonstration of the retinal fundus image before and after preprocessing.

dataset for retinopathy detection consisting of around 88,702 images, in our experiment, we use a very small fraction of it containing images for disease severity grading task with imbalanced classes, requiring us to heavily augment our training data to obtain a model which is stable and not overfitted. The major data augmentation operations that we performed are listed below:

- Rotation: Images were randomly rotated between 0° and 360°
- Shearing: Randomly sheared with angle between 20° and 200°
- Flip: Images were both horizontally and vertically flipped
- Zoom: Images were randomly stretched between (1/1.3, 1.3)
- Crop: Images were randomly cropped to 85%–95% of the original size
- Krizhevsky augmentation: Images were augmented by Krizhevsky color augmentation technique [16]
- Translation: Images were randomly shifted between –25 and 25 pixels,

Also, we scaled and centered each image channel (Red Green Blue (RGB)) to get zero mean and unit variance over the dataset. Figure 6.3 shows some of the post-augmented sample images. In our experiment, each training image was subject to maximum 35 different augmentation methods. The output image sizes of our data augmentation pipeline are 512 × 512 and 448 × 448. Table 6.1 shows some of the statistics related to this procedure.

6.3.3 Network architecture

Table 6.2 illustrates the network architecture of our proposed DR detection method. The input layer of the network is 512 × 512. We tried several kernel filters of sizes 3 × 3, 4 × 4, 5 × 5 and found the best result in the kernel size of 4 × 4. Hence, all the convolutional layers of our network have the kernel size of 4 × 4 with united bias and the stride of 1 except the first and third convolutional layer which have stride of 2.

LeakyReLU [17] was used in all convolutional layer as the activation function for nonlinearity. All the max-pooling layers used have same kernel

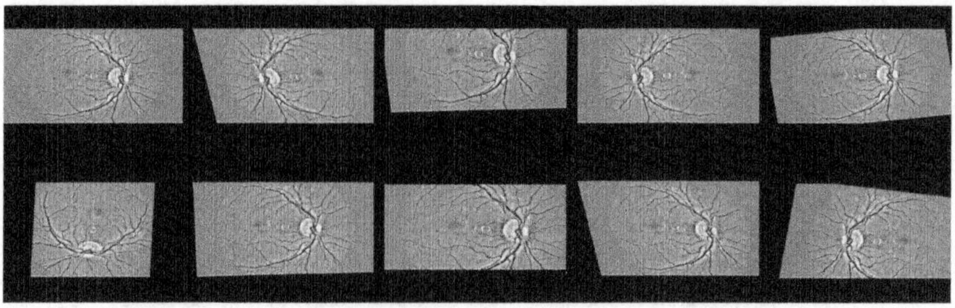

Figure 6.3 Examples of some augmentation operation performed on a preprocessed retinalimages.Aftertheoperation,eachaugmentedimageisresizedmaintainingthe aspect ratio.

Table 6.1 Statistics of the augmentation operations performed on the training set of the Kaggle EyePACS dataset

Grade	*Raw*	*Training*	*Validation*	*Operation*	*Total*
Normal	25,810	25,610	200	0	25,810
Mild NPDR	2,443	2,243	200	11	26,916
Moderate NPDR	5,292	5,092	200	4	25,460
Severe NPDR	873	673	200	27	18,844
Proliferative DR	708	508	200	35	18,288
Total	35,126	34,126	1,000		115,318

Note: Due to highly imbalanced nature of the dataset, different grades were augmented differently.

size of 3×3. The final extracted local features were flattened before passing through fully connected layers. There are two fully connected layers, each having 1,024 neurons. A dropout of 0.5 was added after all but the last fully connected layers to reduce overfitting. Since it is much worse to misclassify severe NPDR or PDR as normal eye than as moderate retinopathy, we considered this multi-class classification as a regression problem and so an output layer of one neuron was added. We took mean squared error as our objective function. Also, we clipped the value output neuron between 0 and 4 since our class ranges between these values.

In the DR experiment, blending the features for both eyes of a patient usually leads to a significant improvement in performance [3]. Thus, we blended our features according to the state-of-the-art blending method [3], where the output of our last max-pooling layer was also used as input features to the blending network, as shown in Table 6.3. To improve the feature quality, feature extraction was repeated as many as 40 times with different augmentations per image. The mean and standard deviation of each feature were used as input to our blending network.

Table 6.2 The proposed network architecture of our early-stage detection and severity grading model

Leyer type	*Kernel size and number*	*Stride*	*Output shape*
Input	-	-	(512,512,3)
Convolution	4 × 4 × 32	2	(256,256,32)
Convolution	4 × 4 × 32	1	(255,255,32)
Max-pooling	3 × 3	2	(127,127,32)
Convolution	4 × 4 × 64	2	(62,62,64)
Convolution	4 × 4 × 64	1	(63,63,64)
Max-pooling	3 × 3	2	(31,31,64)
Convolution	4 × 4 × 128	1	(32,32,128)
Convolution	4 × 4 × 128	1	(33,33,128)
Max-pooling	3 × 3	2	(16,16,128)
Convolution	4 × 4 × 256	1	(17,17,256)
Max-pooling	3 × 3	2	(8,8,256)
Convolution	4 × 4 × 384	1	(9,9,384)
Max-pooling	3 × 3	2	(4,4,384)
Convolution	4 × 4 × 512	1	(5,5,512)
Max-pooling	3 × 3	2	(2,2,512)
Fully connected	1,024	-	(1,024)
Fully connected	1,024	-	(1,024)
Fully connected	1	-	(1)

Note: The depth of the network is 18 layers, while the kernel size of the convolutional filter is 4 × 4. A max-pooling of 3 × 3 is used to downsample the activation map. After convolution layers, two fully connected layers of 1,024 neuron followed by a single neuron output layer are added considering the early-stage detection and severity grading task as a regression problem.

Table 6.3 Network architecture of features blending network

Layer type	*Output shape*
Input layer	(4,096)
Fully connected	(32)
Maxout network	(16)
Fully connected	(32)
Maxout network	(16)
Fully connected	(1)

6.3.4 Training

Our 18-layer-deep proposed network has more than 8.9 million parameters which were randomly initialized using the Xavier weight initialization method. The network was trained with an stochastic Gradient

Table 6.4 Hyperparameters setting of our proposed network architecture

Hyperparameter	*Value*
Objective function	Mean squared error (MSE)
Optimizer	SGD
Momentum	0.9
Multiple learning rates	10^{-4} (for 80 epochs) 10^{-5} (for next 70 epochs) 5×10^{-5} (for next 40 epochs) 10^{-6} (for next 110 epochs)
Batch size	16
Epoch	300

function (SGD) optimization function with 0.90 Nesterov momentum with a fixed schedule over 300 epochs with data augmentation at each step. L2 regularization with a factor of 0.0005 was applied to all weighted layers. We tested several learning rates but found 10^{-4} to be the best initial learning rate. The learning rate constantly decreased as the number of training epochs progressed and ended up having a learning rate of 10^{-6}. The summary of the settings of our training hyperparameters are given in Table 6.4.

The blending network was trained with the Adam [14] optimization algorithm with a fixed schedule over 100 epochs. ReLU activation function was used after each fully connected layer, and an L2 regularization with a factor of 0.001 was applied to every layer. The batch size used was 32 considering mean squared error as our objective function.

6.4 EXPERIMENTAL RESULTS

6.4.1 Dataset

There are several publicly available databases of fundus images for DR including DIARETDB0 [13], DIARETDB1 [12], Kaggle EyePACS [11], and Messidor [7] Databases. DIARETDB0 dataset contains 130 color fundus images, out of which 20 are normal and 110 are affected by DR. On the other hand, DIARETDB1 contains a total of 89 color fundus images, out of which 84 images contain the sign of MAs and 5 images are normal. The Messidor database contains 1,200 retinal fundus images.

In this work, we used EyePACS, the largest publicly available dataset for DR, from the Kaggle Diabetic Retinopathy Detection competition [11].

Figure 6.4. Example of some poor quality retinal images of the Kaggle EyePACS dataset including too bright, dark, and blurred images which make it difficult for the learning algorithm to accurately classify the DR grades.

It is sponsored by the California Healthcare Foundation, which contains a total of 88,702 high-resolution retinal fundus images, captured under a variety of imaging conditions. The dataset also contains artifacts, out of focus, too bright, and too dark images, as illustrated in Figure 6.4.

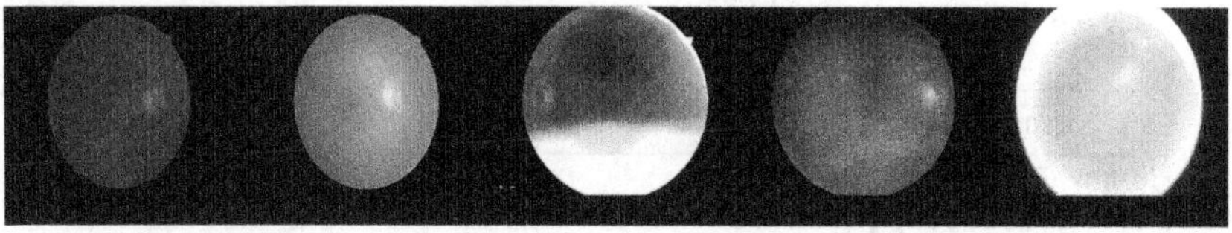

Figure 6.4 Example of some poor-quality retinal images of the Kaggle EyePACS dataset including too bright, dark, and blurred images which make it difficult for the learning algorithm to accurately classify the DR grades.

Table 6.5 Grade distribution in the training set of the Kaggle EyePACS dataset

DR grade	*Grade name*	*Total images*	*Percentage*
0	Normal	25,810	73.84
1	Mild NPDR	2,443	6.96
2	Moderate NPDR	5,292	15.07
3	Severe NPDR	873	2.43
4	Proliferative DR	708	2.01

Fundus images are categorized into five grades (0, 1, 2, 3, 4) according to the severity of DR which are named sequentially as healthy or normal image, mild NPDR, moderate NPDR, severe NPDR, and proliferative DR. The dataset is highly imbalanced due to the presence of around 75% grade 0 (no DR) images. Table 6.5 shows overview of the distribution among different grades in the training set images of the dataset. We splitted the dataset with 35,126 images in the training set and 53,576 images in the test set, as suggested by the Kaggle competition [11] dataset settings. We also used 96% of images of each class in the training set for training and 4% for validation.

6.4.2 Performance evaluation on early-stage detection

In this work, we performed two binary classification for early-stage detection experiment: sick (grades 1, 2, 3, 4) vs healthy (grade 0), and low (grades 0, 1) vs high (grades 2, 3, 4). We considered both of these subproblems equally important for early-stage detection.

We calculated sensitivity and specificity metrics for both of these binary classification problems and found out higher performance in the low–high DR classification than in healthy–sick classification. This is because retinal features of grade (0, 1) are similar to grade (2, 3, 4). Table 6.6 demonstrates the proposed method performance on both classification problems.

6.4.3 Performance evaluation on severity grading

Quadratic weighted kappa, the state-of-the-art performance metric for multi-class classification and suggested evaluation metric for DR [11], is

Table 6.6 Performance evaluation of our proposed method on DR early-stage detection

Classification problem	*Sensitivity (%)*	*Specificity (%)*
Healthy (0) vs Sick (1,2,3,4)	94.5	90.2
Low (0,1) vs High (2,3,4)	98	94

Note: Considering early-stage detection as a binary classification problem, our proposed method achieved astonishing 98% sensitivity and 94% in low–high DR detection.

Table 6.7 Performance evaluation of our proposed method on severity grading of diabetic retinopathy

Metrics	*Score*
Quadratic weighted kappa	0.851
Area under the ROC curve	0.844
F-Score	0.743

adopted as the performance metric of our severity grading prediction. We took thresholds (0.5, 1.5, 2.5, 3.5) to discretize the predicted regression values and make the class levels into integer for computing the Kappa scores. We achieved 0.851 quadratic weighted kappa on the test set of the Kaggle dataset after submitting our solution in Ref. [11].

We also calculated Area Under the ROC Curve (AUROC) and F-Score of our proposed architecture on the same dataset and achieved scores 0.844 and 0.743, respectively. Table 6.7 shows the performance of our proposed method on severity grading of DR on three state-of-the-art evaluation metric.

6.4.4 Comparison among other methods on severity grading

We compare our proposed method with other state-of-the-art methods on severity grading of the Kaggle EyePACS dataset, as shown in Table 6.8. It has been observed that our proposed method exhibits better performance compared to earlier methods proposed in the Kaggle completion [11] and in the literature.

6.5 CONCLUSIONS

In this chapter, we have presented a novel CNN-based deep neural network to detect early-stage and classify severity grades of DR in retinal fundus images. In our work, we have found that without heavy data augmentation, a high capacity network can easily overfit the training data. Even with data augmentation, any network can overfit on oversampled classes such as healthy eye (grade 0). Thus, designing a small capacity network with L2 regularization, and dropout has significant importance in retinopathy

Table 6.8 Comparison with other state-of-the-art methods on severity grading of the Kaggle EyePACS dataset

Methods	*Quadratic weighted kappa*
Min-Pooling [10]	0.84957
o_O [3]	0.84478
Wang et al. [26]	0.84120
Proposed	0.85100

Note: Our proposed network achieves the highest quadratic weighted kappa score of 0.851 compared to other methods proposed in the Kaggle competition and in literature.

detection. So, in this work, we have presented a 4 × 4 kernel-based CNN network with several preprocessing and augmentation methods to improve the performance of the architecture. Our network achieved 98% sensitivity and more than 94% specificity in early-stage detection and a kappa score of more than 0.85 in severity grading on the challenging Kaggle EyePACS dataset. The experimental results have demonstrated the effectiveness of our proposed algorithm to be good enough to be employed in clinical applications.

REFERENCES

1. Acharya, U.-R., Lim, C.-M., Ng, E.-Y.-K., Chee, C., Tamura, T. (2009). Computer-based detection of diabetes retinopathy stages using digital fundus images. *Proceedings of the Institution of Mechanical Engineers, Part H: Journal of Engineering in Medicine*, 223: 545–553.
2. Antal, B., Hajdu, A. (2012). An ensemble-based system for microaneurysm detection and diabetic retinopathy grading. *IEEE Transactions on Biomedical Engineering*, 59: 1720.
3. Antony, M., Brggemann, S. (2015). Kaggle Diabetic Retinopathy Detection Team o O solution.
4. Bravo, M.-A., Arbelez, P.A. (2017). Automatic diabetic retinopathy classification. In: *13th International Conference on Medical Information Processing and Analysis*, 10572: 105721E.
5. Butterworth, D.T., Mukherjee, S., Sharma, M. (2016). Ensemble Learning for Detection of Diabetic Retinopathy. In: *30th Conference on Neural Information Processing Systems (NIPS)*, Barcelona, Spain.
6. Congdon, N.-G., Friedman, D.-S., Lietman, T. (2003). Important causes of visual impairment in the world today. *JAMA*, 290: 2057–2060.
7. Decencire, E., Zhang, X., Cazuguel, G., Lay, B., Cochener, B., Trone, C., Gain, P., Ordonez, R., Massin, P., Erginay, A., Charton, B. (2014). Feedback on a publicly distributed image database: The Messidor database. *Image Analysis & Stereology*, 33: 231–234.
8. He, K., Zhang, X., Ren, S., Sun, J. (2016). Identity mappings in deep residual networks. In: *European Conference on Computer Vision (ECCV)*, Springer, Cham, 630–645.
9. Huang, G., Liu, Z., Van Der Maaten, L., Weinberger, K.-Q. (2017). Densely connected convolutional networks. In: *Proceedings of the IEEE Conference on Computer Vision and Pattern Recognition (CVPR)*, 2261–2269.

10. Graham, B. (2015). Kaggle diabetic retinopathy detection competition report, University of Warwick.
11. Kaggle: Diabetic Retinopathy Detection (2015). https://www.kaggle.com/c/diabeticretinopathy-detection.
12. Kamarainen, T.K.K.K., Sorri, L., Pietil, A.R.V., Uusitalo, H.K. (2007). The DIARETDB1 diabetic retinopathy database and evaluation protocol. In: *Proceedings of British Machine Vision Conference*: 15.115.10.
13. Kauppi, T., Kalesnykiene, V., Kamarainen, J.K., Lensu, L., Sorri, I., Uusitalo, H., Klviinen, H., Pietil, J. (2006). Evaluation database and methodology for diabetic retinopathy algorithms. Machine Vision and Pattern Recognition Research Group, Lappeenranta University of Technology, Finland.
14. Kingma, D.P., Ba, J. (2014). Adam: A method for stochastic optimization. arXiv preprint arXiv:1412.6980.
15. Kori, A., Chennamsetty, S.-S., Alex, V. (2018). Ensemble of convolutional neural networks for automatic grading of diabetic retinopathy and macular edema. arXiv preprint arXiv:1809.04228.
16. Krizhevsky, A., Sutskever, I., Hinton, G.E. (2012). Imagenet classification with deep convolutional neural networks. In: *Advances in Neural Information Processing Systems*, 1097–1105.
17. Maas, A.L., Hannun, A.Y, Ng, A.Y. (2013). Rectifier nonlinearities improve neural network acoustic models. In: *Proceedings of ICML*, 30: 3.
18. Melville, A., et al. (2000). Complications of diabetes: Screening for retinopathy and management of foot ulcers. *Quality and Safety in Health Care*, 9: 137–141.
19. Pratt, H., Coenen, F., Broadbent, D. M., Harding, S. P., Zheng, Y. (2016). Convolutional neural networks for diabetic retinopathy. *Procedia Computer Science*, 90, 200–205.
20. Roychowdhury, S., Koozekanani, D.D., Parhi, K.K. (2014). Diabetic retinopathy analysis using machine learning. *IEEE Journal of Biomedical and Health Informatics* 18: 1717–1728.
21. Silberman, N., Ahrlich, K., Fergus, R., Subramanian, L. (2010, March). Case for automated detection of diabetic retinopathy. In *2010 AAAI Spring Symposium Series.*
22. Simonyan, K., Zisserman, A. (2014). Very deep convolutional networks for large-scale image recognition. arXiv preprint arXiv:1409.1556.
23. Sopharak, A., Uyyanonvara, B. and Barman, S. (2009). Automatic exudate detection from non-dilated diabetic retinopathy retinal images using fuzzy c-means clustering. *Sensors*, 9: 2148–2161.
24. Szegedy, C., Ioffe, S., Vanhoucke, V., Alemi, A.A. (2017). Inception-v4, inception-resnet and the impact of residual connections on learning. *AAAI*, 4: 12.
25. de la Torre, J., Valls, A., Puig, D. (2017). A deep learning interpretable classifier for diabetic retinopathy disease grading. arXiv preprint arXiv:1712.08107.
26. Wang, Z., Yang, J. (2017). Diabetic retinopathy detection via deep convolutional networks for discriminative localization and visual explanation. arXiv preprint arXiv:1703.10757.
27. Wang, S., Yin, Y., Cao, G., Wei, B., Zheng, Y., Yang, G. (2015). Hierarchical retinal blood vessel segmentation based on feature and ensemble learning. *Neurocomputing*, 149: 708–717.
28. Yang, Y., Li, T., Li, W., Wu, H., Fan, W., Zhang, W. (2017). Lesion detection and grading of diabetic retinopathy via two-stages deep convolutional neural networks. In: *International Conference on Medical Image Computing and Computer-Assisted Intervention*, 533–540.

Chapter 7

Vehicle detection using faster R-CNN

Kaustubh V. Sakhare, Pallavi B. Mote, and Vibha Vyas
SPPU University

CONTENTS

7.1 INTRODUCTION

Object detection works as an exacting predicament in the domain of computer vision. The necessity of addressing it within great dynamic conditions has attracted a lot of research interest in this area. Vehicle detection is essential in various applications ranging from traffic surveillance, public safety, and security up to autonomous driving. Object detection in a controlled environment has remarkably minimized detection rate and accuracy score. The initial efforts are focused on the extraction of histogram of gradient (HOG), scale invariant feature transform (SIFT), and Harr-like approach [1] as hand-crafted features. However, it is a challenging task to detect vehicles under different pose, camera viewpoint, scale, occlusion, lightning, and weather conditions [2]. In recent years, there is an explosion of deep learning-based approaches which perform great in generic object detection. Various deep learning models perform with the availability of extensive databases in the multiclass classification. However, with the same model, there is no surety of achieving exceptional performance for a particular application such as vehicle detection in autonomous vehicles [15]. In this context, region-based convolutional neural network (CNN) formulates fast R-CNN.

Incorporation of selective search algorithms as region proposals gives a decisive breakthrough in achieving real-time object detection if we overlook the time required to find out the region proposals [16]. Moving ahead, region proposals, along with sharing of the convolutional layers [1,2] popularly phrased as region proposal networks (RPNs), result in reduced computational complexity. Faster R-CNN, use object proposal methods to propose candidate regions to the classifier, evolved as the top performers for object detection on standard benchmark datasets. RPN acts as a fully connected network (FCN). It is constructed by appending convolutional layers capable of regressing the region's bounds and objectness scores. This training generates detection proposals. The selective search algorithm is employed here to extract 2,000 proposals from a single image, followed by region of interest (RoI) pooling to extract the fixed size of the feature map from each proposal. L1 loss with softmax loss is combined to characterize the classification probability. Faster R-CNN expressed lowered testing time compared to the fast R-CNN and comparable mean average precision [4,5,17].

Faster R-CNN is used to speed up the object detection mechanism. The performance is compared with the single-shot detector (SSD) for object detection. The significant contributions of this work involved:

1. After receiving the input from the convolution layers, RPN generates region proposals. For accurate proposals, bounding box regression correction anchors are used.
2. An activation function is used to separate anchors to the foreground and background.
3. The MIO-TCD [12] dataset containing 650,000 low-resolution images, which are divided into 11 categories, is used to validate the performance of proposed faster R-CNN architecture with SSD.
4. L1 loss with softmax loss is used together to characterize the classification probability.

The organization of the paper is organized as: related work is mentioned in Section 7.2. Vehicle detection and overview of faster R-CNN are presented in Section 7.3. Experimentation and result analysis are detailed in Section 7.4. The conclusion and application of the proposed algorithm are discussed in Section 7.5. The subsequent section maintains the references for the article.

7.2 RELATED WORK

Various approaches are involved in the vehicle detection methods, which include motion, hand-crafted feature, and CNN-based approaches. Success in classification and detection of the vehicle using hand-crafted

feature-based approach depend upon the skill test of the programmer and cannot meet the optimum feature representation [7]. In the context of vehicle detection, hand-crafted features are extracted for the learning model. The object detection and localization methods are analyzed based on the quality of the feature extraction. The shape is one of the prominent features used to represent the object. Few driving factors for selection of the shape descriptors are it should be invariant to the translation, rotation, and scaling. Shape descriptors should be invested insightfully in object detection [2,15]. A large number of techniques have been proposed for describing shapes in object detection wherein the points of interest are discerned out of the images and compared to those with the ones registered from dataset images to find the object of interest [2]. This part of interest inside the image is normally treated as the feature. Compared to other detectors, speeded up robust features (SURF) and SIFT feature detectors are robust. Even a better localization quality as expected than others while considering real-time vehicle detection. It quickly detects and classifies the objects as these are rotation and scale-invariant. These methods are employed in feature extraction for a long time.

Enormous work has been carried out in the feature detector approach and is summarized in Table 7.1 [2,13].

The various classification techniques and learning algorithms ranging from conventional machine learning techniques up to deep neural networks are explored for its usage on vehicle detection. Selecting features is a crucial task to achieve high accuracy. The survey by Cheng et al. [1,2,15] details the conventional methods used in object detection based on features such as HOG, Texture, and Bag of Words and classifiers such as k-Nearest Neighbor,

Table 7.1 Performance of feature detector algorithm

Feature detector	*Invariance*			*Qualities*			
	Rotation	*Scale*	*Affine*	*Repeatability*	*Localization*	*Robustness*	*Efficiency*
SURF	Yes	Yes	No	Intermediate	Strong	Intermediate	Strong
SIFT	Yes	Yes	No	Intermediate	Strong	Strong	Intermediate
Harris	Yes	No	No	Strong	Strong	Strong	Intermediate
Harris Laplace	Yes	Yes	No	Strong	Strong	Intermediate	Weak
Hessian	Yes	No	No	Intermediate	Intermediate	Intermediate	Weak
Salient Regions	Yes	Yes	Yes	Weak	Weak	Intermediate	Weak
SUSAN	Yes	No	No	Intermediate	Intermediate	Intermediate	Strong
MSER	Yes	Yes	Yes	Strong	Strong	Intermediate	Strong
Hessian Laplace	Yes	Yes	No	Strong	Strong	Strong	Weak
DoG	Yes	Yes	No	Intermediate	Intermediate	Intermediate	Intermediate

Artificial Neural Network, and Support Vector Machines (SVMs), with a sliding window technique. The work carried out in Refs [14,16] recites the performance of hand-crafted features with machine learning algorithms. The typical challenge in all these proposed methods is it depends on the features derived from the instances, and the performance may be dropped when extended to the more complex problems. The human ingenuity in the feature extraction has led to the emergence of CNN-based methods in the field of object detection. The CNN-based approaches can be categorized as pre-trained and hybrid CNN models [6]. CNNs accepting the object proposals are seen as one of the breakthroughs in compact feature representation. A comprehensive work can be seen on object proposal methods such as selective search [7], other algorithms based on grouping of super pixels [17], and sliding window-based algorithms [8,9]. Object proposal methods, along with detectors, are seen in R-CNN and fast R-CNN [18].

R-CNN trains CNNs classifying the proposal into object categories or background. The accuracy of the model depends on the region proposal modules. A lot of prior work uses deep learning algorithms to predict bounding boxes [10]. Fully connected layers detect multiple class-specific objects. Multibox proposal networks are applied to a single image crop or multiple large image crops [10]. It doesn't share convolution features between detectors and proposals. Fast R-CNN [2] facilitates the end-to-end detector on shared convolutional features and demonstrates extensive accuracy.

7.3 VEHICLE DETECTION USING FASTER R-CNN

The methods used for object detection detect general object classes, but vehicle detection is specific class object detection. The set of frameworks suitable for the generic object detection does not perform up to the mark for specific class object detection. To improvise the accuracy of those models, one of the ways is training a network with database of specific class categories. But such approach does not guarantee performance improvement. Using the default parameter and trained with the KITTI car dataset, the faster R-CNN model can perform reasonably well when compared to top challengers in the KITTI car competition [4]. The performance of faster R-CNN is reasonably found on training and test scaling [4]. Also, specifically for KITTI dataset, performance can be improved using the iterative approach. Table 7.2 shows various pre-trained CNN architecture models and the dataset on which they are trained.

7.3.1 Faster R-CNN outline

Faster R-CNN has significantly achieved the reduction in the detection time. RPNs are combined together to form region proposals using CNN preferred over the selective search algorithm. Apart from improvement in

Table 7.2 Pre-trained models

Model name	*Dataset on which it is pre-trained*
VGG16	ImageNet
VGG19	ImageNet
Inception V3	Kaggle
MobileNet	ImageNet
Mask R-CNN	MS COCO
YOLOv2	Raccoon dataset

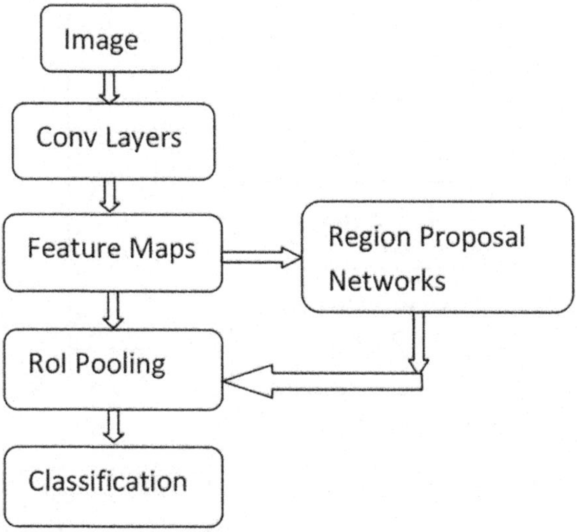

Figure 7.1 Faster R-CNN network architecture.

the quality box proposals, numbers of proposals are minimized to almost 25% of their original size. Faster R-CNN, represented in Figure 7.1, shares a fully connected layer between the object classifier and RPN. The mechanism detects the most suitable proposals from a wide range of aspect ratios and scales. These layers are trained jointly in an effective manner.

The faster R-CNN stepwise approach is mentioned as follows:

1. **Convolution Layer**: Feature maps are generated using convolution layers, which include convolution, ReLU, and pooling operations. Reset 101 and Inception-v2 are the pre-trained models used in faster R-CNN and SSD architecture [8,9,10].
2. **Region Proposal Network (RPN)**: RPN generates region proposals; anchors are used for generating accurate proposals. The softmax layer determines whether the anchor belongs to the foreground or

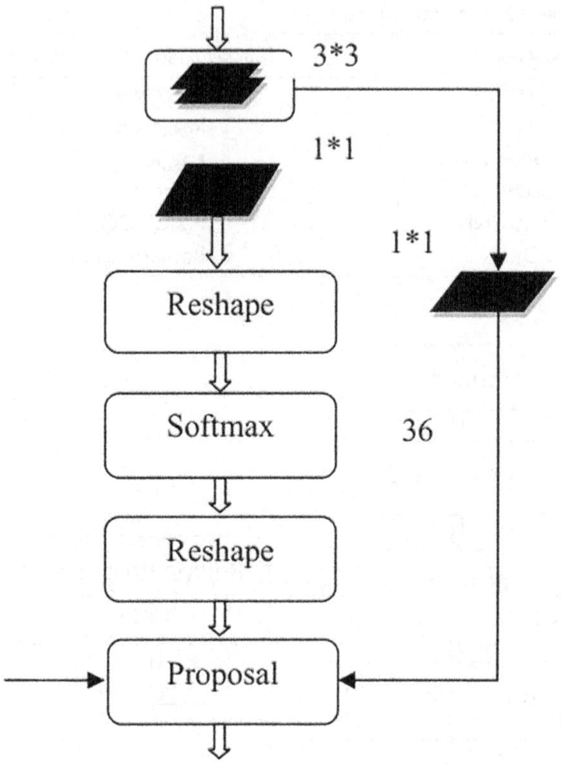

Figure 7.2 RPN structure.

background. The anchor box functions as a reference point to different region proposals centered on the same pixel. Collectively RPN is responsible for detection of time reduction at a great level. Figure 7.2 shows the RPN network, which works through two branches. It gets foreground or background through the first branch, and with the help of the second branch, bounding box regression offset for anchors is calculated to get accurate proposals. The proposal layer excludes very small proposals, beyond boundaries proposals, and selects the foreground anchors with accurate bounding box regression offset [10].

3. **RoI Pooling Layer:** Proposed regions after RPN of different sizes result in different sized feature maps. The RoI pooling layer reduces the feature maps into the same sizes by collecting proposals and input feature maps. It significantly speeds up both train and test time.
4. **Classification:** Fixed-size feature maps are used to calculate the class of proposal by predicting class probabilities through a fully connected layer and a softmax layer. Bounding box offset prediction is obtained using bounding box regression and achieves an accurate target detection box. Figure 7.3 shows the classification layer structure [10].

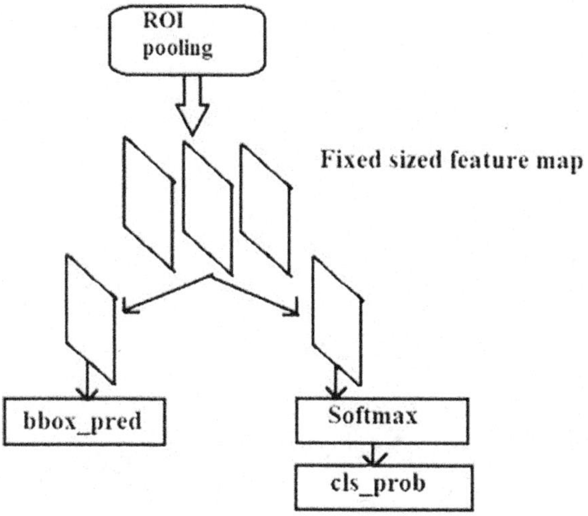

Figure 7.3 Classification layer structure.

7.4 EXPERIMENTS AND RESULT ANALYSIS

This section details the experimental setup used during the training and validation. Faster R-CNN and SSD algorithms are validated on the MIO-TCD database. The model performance is measured based on the object detection rate and object recognition rate. The subsequent discussion in this section takes a closer look at the database, experiments, and result analysis.

7.4.1 Training dataset

The MIO-TCD [13] dataset aims to measure the performance of various object classification and localization algorithms. It contains around 650,000 low-resolution images, which are divided into 11 categories: Pedestrian, Bus, Articulated Truck, Bicycle, Car, Pickup Truck, Motorcycle, Non-Motorized Vehicle, Single Unit Truck, Background, and Work Van. It is partitioned as 80% training images and 20% testing images. A background category is constructed by a random sampling of 200,000 images. This dataset has acceptable variations for training a real-time traffic monitoring system. The images are taken at different times of the day from various angles, resolutions, and scales [13].

Two architectures are used for the validation as faster R-CNN and SSD [9,10]. Resenet101 model pre-trained on the Image Net database is used as the base model in the implementation of faster R-CNN. The Inception-v2 model pre-trained on the ImageNet database is used as the base CNN

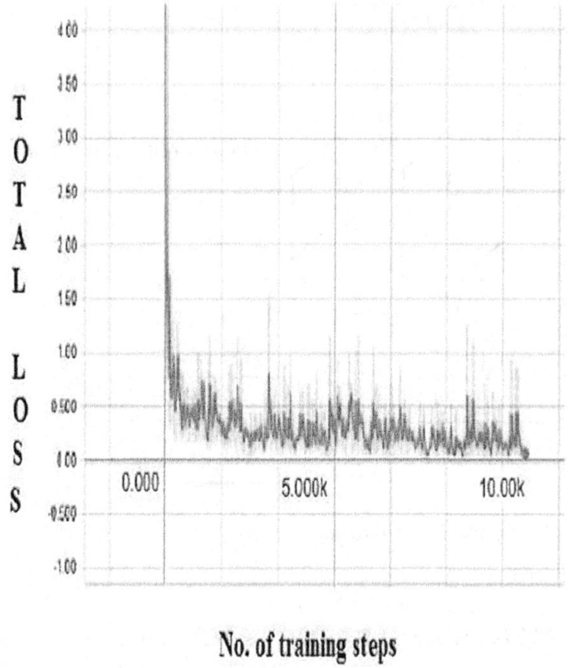

Figure 7.4 Total losses of faster R-CNN.

model in the SSD architecture. Both the networks are trained on the five classes as the bicycle, bus, car, motorcycle, and single-unit truck [8,9,10]. 10% of the images are kept for testing from the selected dataset of each class with 400 training images. Figures 7.4 and 7.5 show the total losses during the training of faster R-CNN and SSD architectures, respectively.

From the above results, we conclude that using pre-trained networks, training losses reduce near to zero with small datasets as well. Object detection and recognition rate of vehicle detection are computed, as shown in Figures 7.6 and 7.7, respectively, for both faster R-CNN and SSD architectures [3,10].

$$\text{Object detection rate} = \frac{\text{Total number of detected objects} \times 100}{\text{Total number of objects}}$$

The results for the same are as follows.

The objectness score is calculated, which shows how likely for an image window to contain an object of any class such as bicycle, car in the database when compared to backgrounds such as roads and buildings. The objectness score of each class for both the architectures is shown in Figure 7.8.

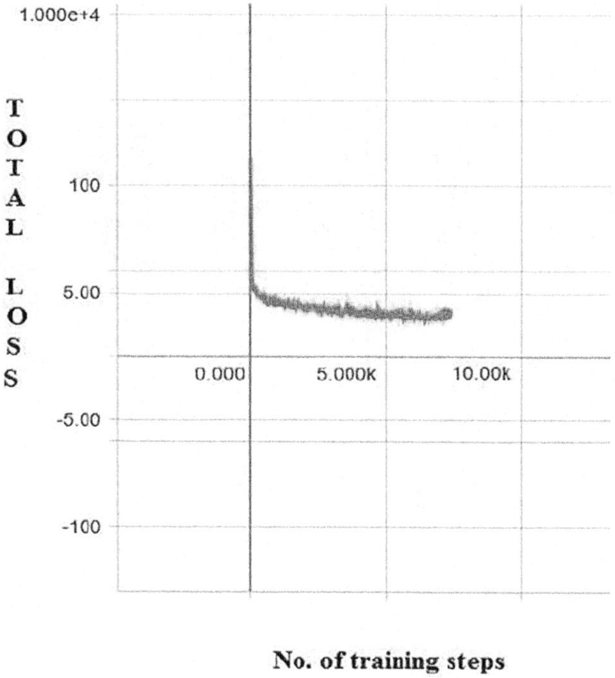

Figure 7.5 Total losses of SSD.

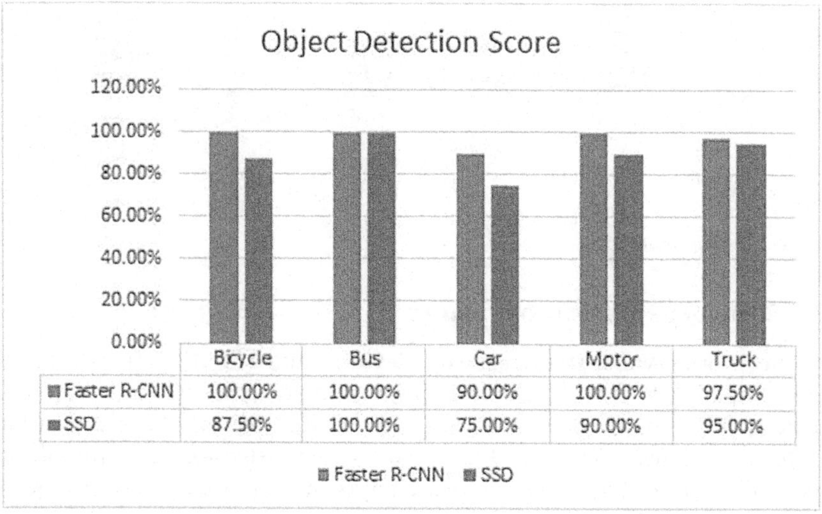

Figure 7.6 Object detection rate.

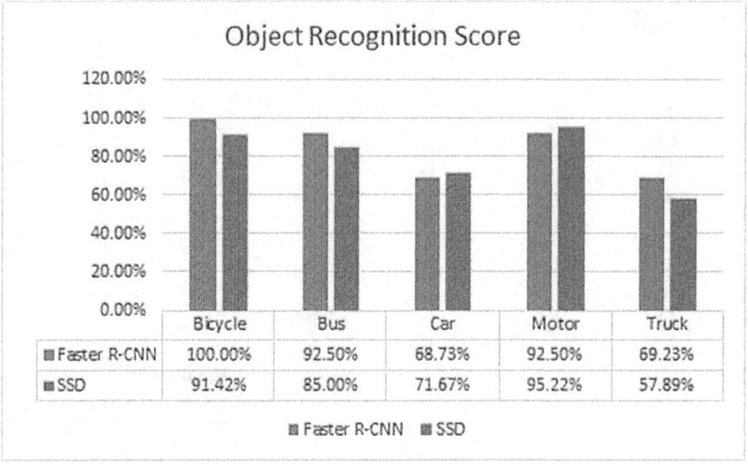

Figure 7.7 Object recognition rate.

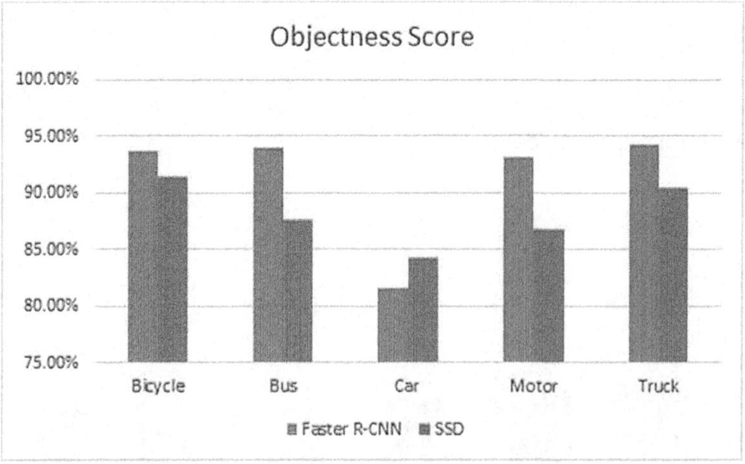

Figure 7.8 Objectness score.

7.4.2 Interpretation of results

In this work, two networks, faster R-CNN and SSD [11] architectures, are compared. SSD is the single-shot multibox detector which uses multi-scale anchor boxes for the detection and classification task. The evaluation factor considered is the objectness score, object detection rate, and object recognition rate. Experimentations comprising the listed class of objects have been carried out on a dynamic selection of images from the MIO-TCD database.

- Faster R-CNN achieves a higher object detection rate than SSD.
- Comparatively, faster R-CNN achieves better object recognition rate than SSD though the object recognition rate for class car and motorcycle of SSD is slightly higher than faster R-CNN.
- Comparatively, faster R-CNN has more objectness score for the class identified by both the architectures.
- However, multi-scale anchor boxes in SSD are found to be improvised along with non-maxima suppression outperforming the faster R-CNN on computational complexity.

Validation of the proposed techniques on the augmented database is the latitude of the work for generic vehicle detection in driver assistance systems.

7.5 CONCLUSIONS

In this chapter, the faster R-CNN is validated for vehicle detection systems. For better performance in real-world challenges, the experiments are performed using two different training networks on the MIO-TCD dataset with the multiclass classification. The variation in the anchor scales is performed, resulting in the conclusion that the small-scale anchor gives better performance at distant vehicle detection. Objectness score, object detection rate, and object recognition rate are considered as an evaluation index for comparing architectures. Faster R-CNN achieves higher object detection rate and object recognition rate than SSD, an exception to the class like car and motorcycle.

REFERENCES

1. Jon Arrospide, Luis Salgado: "A Study of Feature Combination for Vehicle Detection Based on Image Processing," The *Scientific World Journal*, 2014 (2), 196251, 2014.
2. Dilip K. Prasad: "Survey of the Problem of Object Detection in Real Images," *International Journal of Image Processing (IJIP)*, 6 (6), 441–466, 2012.
3. Anitya Kumar Gupta: "Time Portability Evaluation of RCNN Technique of OD Object Detection–Machine Learning (Artificial Intelligence)," *International Conference on Energy, Communication, Data Analytic and Soft Computing (ICECDS-2017)*, 2017.
4. Quanfu Fan, Lisa Brown, John Smith: "A Closer Look at Faster R-CNN for Vehicle Detection," *IEEE Intelligent Vehicles Symposium (IV) Gothenburg, Sweden*, June 19–22, 2016.
5. R. Girshick, J. Donahue, T. Darrell, J. Malik: "Rich Feature Hierarchies for Accurate Object Detection and Semantic Segmentation," *In Proceedings of the IEEE Conference on Computer Vision and Pattern Recognition*, pp. 580–587, 2014.
6. Shih-Chung Hsu, Chung-Lin Huang, Cheng-Hung Chuang: "Vehicle Detection Using Simplified Fast R-CNN," 2018.

7. Chen Linkai, Feiyue Ye, Yaduan Ruan, Honghui Fan, Qimei Chen: "An Algorithm for Highway Vehicle Detection Based on Convolutional Neural Network."*EURASIP Journal on Image and Video Processing* 2018 (1), 109, 2018.
8. Chengcheng Ning, Huajun Zhou, Yan Song, Linhui Tang: "Inception Single Shot Multibox Detector for Object Detection," *Proceedings of the IEEE International Conference on Multi-media and Expo Workshops (ICMEW)*, 2017.
9. SSD Architecture, http://medium.com/jonathanhui/ssd-object-detection-single-shotmultibox-detector-for-real-time-processing.
10. Shaoqing Ren, Kaiming He, Ross Girshick, Jian Sun: "Faster R-CNN: Towards Real-Time Object Detection with Region Proposal Networks," arXiv:1506.01497v3 [cs.CV], 2016.
11. Youngwan Lee, Huieun Kim, Eunsoo Park, Xuenan Cui, Hakil Kim: "Wide-Residual-Inception Networks for Real-time Object Detection," *2017 IEEE Intelligent Vehicles Symposium (IV)* June 11–14, Redondo Beach, CA, 2017.
12. Luo Zhiming, Frederic Branchaud-Charron, Carl Lemaire, Janusz Konrad, Shaozi Li, Akshaya Mishra, Andrew Achkar, Justin Eichel, Pierre-Marc Jodoin. "MIO-TCD: A New Benchmark Dataset for Vehicle Classification and Localization." *IEEE Transactions on Image Processing* 27 (10), 5129–5141, 2018.
13. Ehab Salahat, Murad Qasaimeh: "Recent Advances in Features Extraction and Description Algorithms: A Comprehensive Survey," arXiv:1703.06376v1 [cs.CV], 2017.
14. Dalal Navneet, Bill Triggs: "Histograms of Oriented Gradients for Human Detection." In *2005 IEEE Computer Society Conference on Computer Vision and Pattern Recognition (CVPR'05)*, vol. 1, pp. 886–893. IEEE, 2005.
15. Sakhare Kaustubh V., Tanuja Tewari, Vibha Vyas. "Review of Vehicle Detection Systems in Advanced Driver Assistant Systems." *Archives of Computational Methods in Engineering*, 27 (2), 591–610, 2020.
16. Tanuja Tewari, K.V. Sakhare, Vibha Vyas: "Vehicle Detection in Aerial Images using Selective Search with a Simple Deep Learning based Combination Classifier," *Proceedings of the Third International Conference on Microelectronics, Computing and Communication Systems*, 221–233, 2019.
17. Tayara Hilal, Kim Gil Soo, Kil To Chong: "Vehicle Detection and Counting in High-Resolution Aerial Images Using Convolutional Regression Neural Network." *IEEE Access*, 6, 2220–2230, 2017.
18. Hosang Jan, Rodrigo Benenson, Piotr Dollár, Bernt Schiele: "What Makes for Effective Detection Proposals?." *IEEE Transactions on Pattern Analysis and Machine Intelligence*, 38 (4), 814–830, 2015.

Chapter 8

Two phase authentication and VPN-based secured communication for IoT home networks

Md Masuduzzaman and Ashik Mahmud
American International University-Bangladesh

Anik Islam
Kumoh National Institute of Technology

Md Mofijul Islam
University of Dhaka

CONTENTS

8.1 INTRODUCTION

8.1.1 Background

The Internet of Things (IoT) is a system of interconnected devices, sensors, and actuators etc., which work together in a network to reach a common goal [1]. Such technology can be implemented in various ways to make

our daily lives easier by placing internet capabilities in devices which are not regularly used as network devices. In recent years, the application of microprocessor-based controllers in devices ranging from toasters to airliners is being added to connect them to the internet [2–4]. With such advancements in IoT technology, it is showing potential to be deployed as consumer products for home usage [4]. Providing internet capabilities to home devices, such as air conditioners, lights, fans, refrigerators etc., enables them to be controlled remotely. Such type of applications can be called home IoT networks as the devices are connected to a home network, and these devices can be controlled remotely which make these devices vulnerable to malicious attacks [3–5], and secure data transferring is also needed [2]. So there is a need to add methods of authentication and security between the user and the devices in the network in order to prevent attacks [6–8]. Secure authentication of devices in a network includes a key exchange mechanism and handshake between the devices [7–9]. This can be done through a centralized network system in which a central node is responsible for the security mechanisms or through distributed networks where each node shares private keys, and after successful handshake communication is established [7–9].

8.1.2 Authentication protocols

Authenticating devices in a home network is a key process in securing the user interaction with IoT. If the network lacks security, the end devices are vulnerable to attack and the purpose of implementing IoT is diminished [10]. There are several authentication mechanisms for IoT applications. Such as password-based remote user authentication using one-way hash functions and ticket-based authentication [11]. Existing methods of authentication of devices include registering devices in a cloud platform running with the home network and initiating a handshake and key exchange [11,12]. Such systems also include a current method of key exchange which is the RSA key exchange mechanism [7]. As most of the IoT devices are resource constrained, a Datagram Transport Layer Security (DTLS) protocol is often used for handshaking [13].

8.2 RELATED WORKS

8.2.1 Wi-Fi network-based security

The system contains a home gateway (GW), and the user or mobile device and several IoT devices are connected through a Wi-Fi network, as shown in Figure 8.1.

Users can access the IoT devices from the home GW, and the GW performs authentication and monitoring functions between the devices in the system [4]. The authentication protocol used in this system involves the use

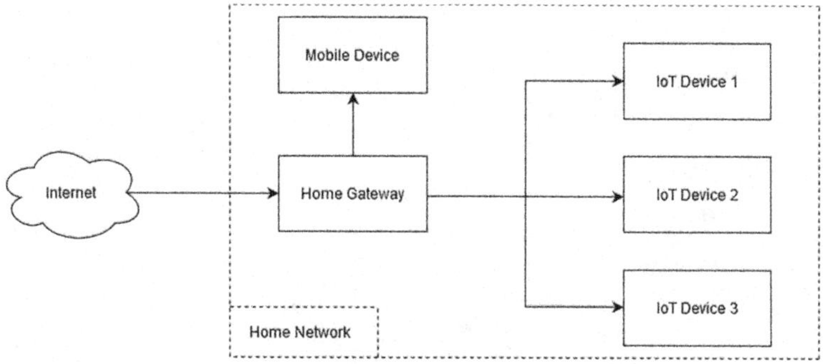

Figure 8.1 System step for Wi-Fi-based security.

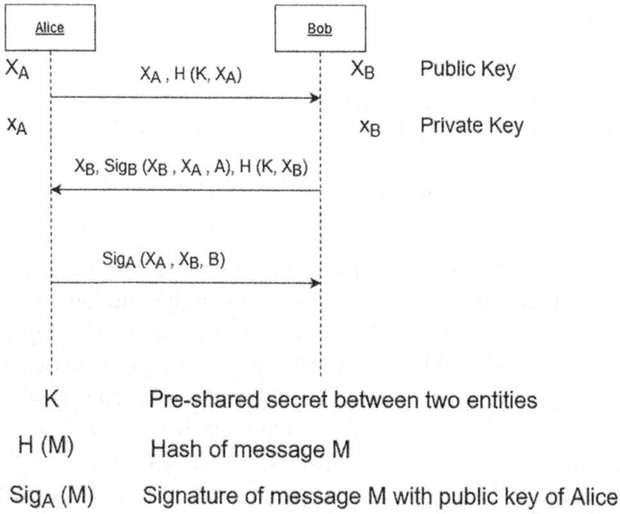

Figure 8.2 Simple handshake protocol using ECC.

of public key cryptography [7] with pre-shared keys between the GW and a new device which utilizes elliptic curve cryptography (ECC) to reduce the key size [4], as shown in Figure 8.2. This model lacks a proper mechanism for the handshaking protocol for constrained devices. Although ECC was used to reduce key size shared for authentication, the whole application lacks security measures if a malicious attacker is able to infiltrate the system.

8.2.2 PAuthKey protocol

The system authenticates its devices in a two-phase authentication protocol [7,9,11], as shown in Figure 8.3. The network consists of several nodes in a cluster that communicates with the user over the IoT cloud through a GW.

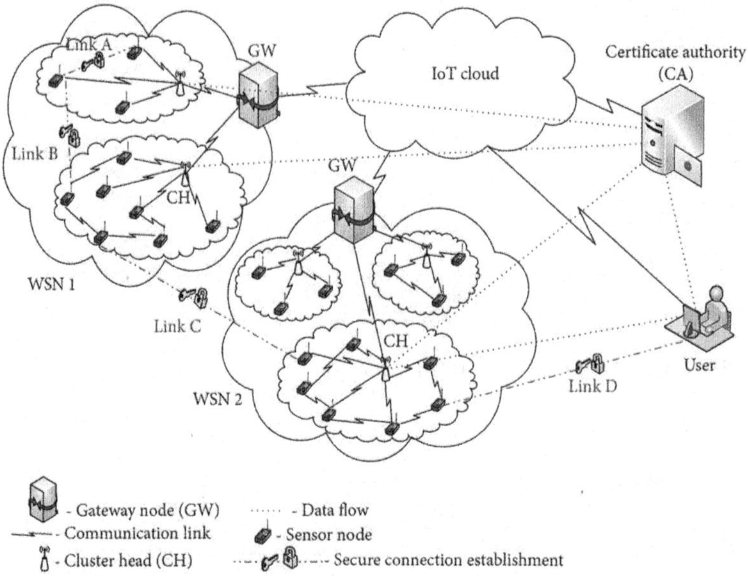

Figure 8.3 PAuthKey network model.

A certificate authority (CA) is connected to the network and is responsible for authenticating the devices using handshake and key exchange. The first phase of the authentication is carried out manually by registering the devices in the CA [9–11]. The second phase is carried out through initiating handshake using DTLS handshaking protocol and exchanging keys using ECC [7,8,13]. Since the user communicates with the IoT devices over the IoT cloud or internet, in these areas, the data packets remain vulnerable to attackers. Even though they are encrypted, there still lie possibilities of the packets being sniffed and decrypted.

8.2.3 Two-way authentication security scheme on existing DTLS protocol

This system comprises a network containing a CA which is responsible for authorizing the devices and an access control server for exchanging key, as shown in Figure 8.4. The devices communicate with each other over the internet and the IoT devices communicate with the user and the server through a GW [14]. The key aspect of this system is the use of DTLS handshake for mutual authentication of the devices, i.e., a two-way authentication [13–16], as shown in Figure 8.5. During the handshake, keys are exchanged using RSA cryptography [14]. This model also briefly talks about using virtual private network (VPN) as a mechanism to secure payload over the internet for their proposed model [10,14,17]. As this model

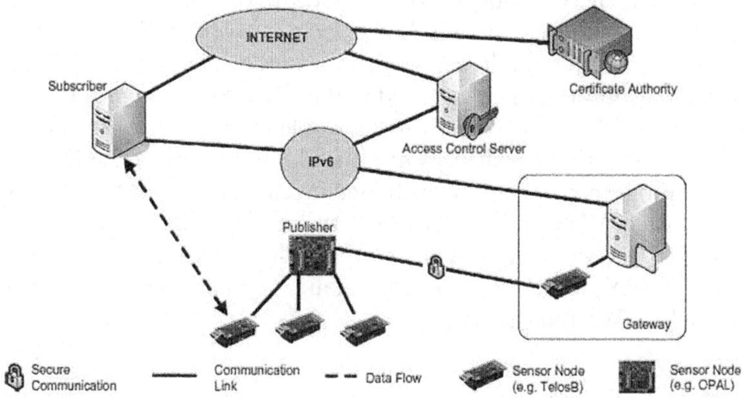

Figure 8.4 DTLS-based security and two-way authentication network model.

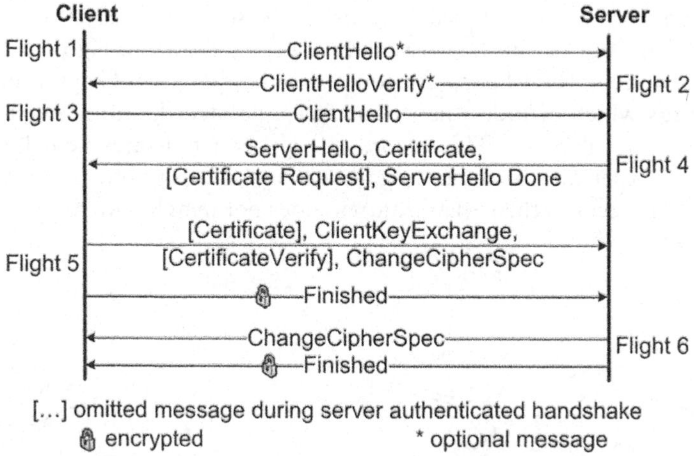

Figure 8.5 DTLS handshake mechanism for two-way authentication.

uses the RSA key exchange mechanism, there exist the probability of large network overheads occurring due to large key size. This can be modified by using a key exchange mechanism which generates a lower key size for the same security bit for RSA.

8.3 NETWORK MODEL AND ASSUMPTION

Our proposed network model follows the existing network model used in the PAuthKey system for the authentication process [11]. Since the devices in their network are communicating over the internet or IoT cloud [11,18],

we modified the network using VPN for secure communication. Here, the model is made such that the wireless sensor network (WSN) clusters with GWs can be an individual house or a room so that the system is scalable to larger implementations. The CA used in the PAuthKey system [11] has been replaced by a VPN server which is responsible for registering the devices as well as authenticating the devices using DTLS handshake [11,14] and public key cryptography for a key exchange mechanism [7]. The VPN server also establishes VPN endpoint to GW tunnels [19,20] for securing data packets sent by any devices in the system. The VPN server is responsible for creating VPN tunnels for data communication between the user and the device or the GWs and the server.

8.4 PROPOSED SOLUTION

8.4.1 MAC address-based user registration

Similar to the aforementioned systems, our system also uses a registration phase for the devices in the network [4,8,14], as shown in Figure 8.6.

In our system, the VPN server contains a database that keeps a record of valid users which includes their usernames, password, and MAC address, as shown in Table 8.1. This record is kept as a fail-safe such that if an authorized user does manage to gain access to the system, the server can deny them access as their MAC address does not match with any registered addresses.

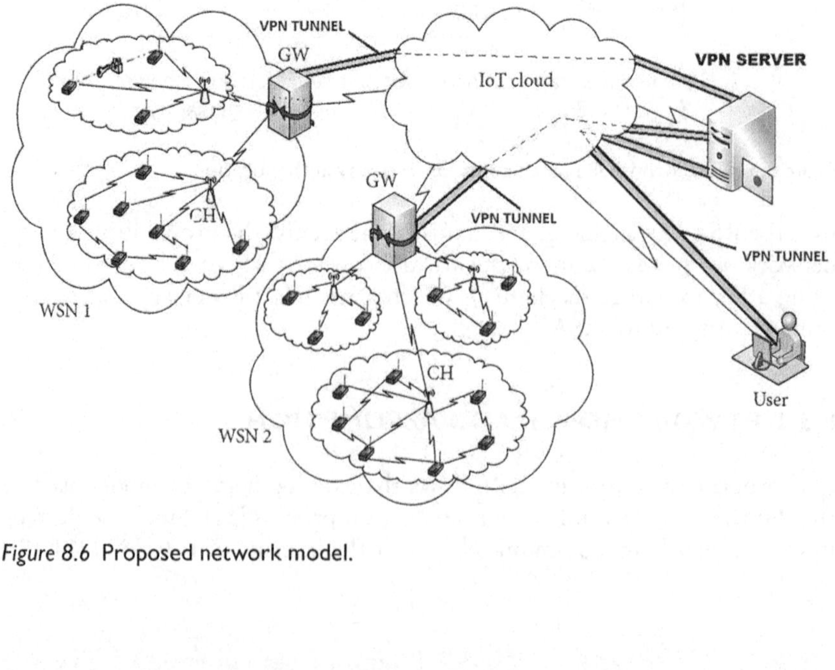

Figure 8.6 Proposed network model.

Table 8.1 MAC address-based user registration

User	*Password*	*MAC*
user1	pass1	3.-65-EC-6F-C4-58
user2	pass2	00-1B-63-84-45-E6
user3	pass3	00-1B-44-113A-B7

Table 8.2 Comparable key sizes between RSA and ECC

Security bits	*ECC*	*RSA*
80	260	1,024
112	224	2,048
128	256	3,072
192	384	7,680
256	521	12,350

8.4.2 Authentication protocol

Use of ECC is now being adopted as an alternate public key exchange mechanism for IoT-based networks [4,21]. In PAuthKey [11] and Wi-Fi-based systems [4], the usage of ECC was also adopted along with DTLS handshake. On the other hand, the DTLS two-way authentication protocol used RSA as its method for key exchange. In our proposed method, we use ECC along with DTLS handshake for authentication purposes. The reason ECC is used in lieu of RSA is that ECC is more suitable for resource-constrained devices. ECC keys are generated through computation on an elliptical curve whose basic equation is

$$y^2 = x^2 + ax + b \tag{8.1}$$

The trapdoor function of ECC is similar to RSA with complex mathematical computation [15]; this is due to elliptic curves having horizontal symmetry and any non-vertical line will intersect the curve in three places at most. Moreover, the endpoint of the non-vertical intersection can be reflected to form another non-vertical intersection, resulting in the scope of more key generation [21,22]. The major advantages that ECC has over RSA are:

ECC relies on difficult discrete logarithm functions which make it more difficult to decrypt by malicious attackers [22].

ECC generates keys with shorter size compared to RSA for the same security bits [21,22].

For the same bit size, ECC generates more number of keys than RSA [22], which is provided in Table 8.2.

8.4.3 Data transfer security using VPN

In our network model and the network models discussed in our related works, users communicate with IoT devices through the internet or IoT cloud [4,11,14]. This may leave the data packets being sent vulnerable to eavesdropping or sniffing attacks. To solve this problem, our model introduces VPN to the network. A VPN endpoint to the GW tunnel [20] is created from the server to the GWs, from user to server, and from user to GW by the VPN server, as shown in Figure 8.4. This enables the devices to communicate using private IPs over a secured channel. Furthermore, such type of secured communication ensures that confidentiality, integrity, and authenticity are maintained in the system.

8.5 CONCLUSIONS AND FUTURE WORK

8.5.1 Scope of future work

This solution focuses on secure data communication and authentication protocols for home IoT networks. The proposed network model mainly focuses on home networks including control over multiple homes for one user based on the cluster topology. This network is scalable to larger network areas such as industries and hospitals and even smart cities. Furthermore, the scope of improving VPN-based communication over the internet is vast as this solution focus on conceptual-based discussion rather than practical implementation.

8.6 CONCLUSIONS

In this solution, the problem that was focused on the authentication mechanism of devices is in a home IoT network and secured communication of devices over the internet. An efficient authenticating mechanism based on ECC and DTLS handshake was introduced, and VPN-based tunneling was proposed to secure data communication. Moreover, MAC address-based fail-safe solution was also proposed such that in an unlikely event an unauthorized user gains access to the system, they can be dealt with.

REFERENCES

1. Parvaneh Asghari, Amir Masoud Rahmani, Hamid Haj Seyyed Javadi, Internet of Things applications: A systematic review. *Computer Networks* 148, 241–261, 2019,
2. Anik Islam, Soo Young Shin, A blockchain-based secure healthcare scheme with the assistance of unmanned aerial vehicle in Internet of Things. *Computers & Electrical Engineering* 84, 106627, 2020, https://doi.org/10.1016/j.compeleceng.2020.10662

3. Huichen Lin, Neil W. Bergmann, "IoT Privacy and Security Challenges for Smart Home Environments," University of Queensland, Australia, 2016.
4. Freddy K. Santoso, Nicholas C. H. Vun, "Securing IoT for Smart Home System," *IEEE International Symposium on Consumer Electronics (ISCE)*, 2015.
5. Shivaji Kulkarni, Shrihari Durg, Nalini Iyer, "Internet of Things (IoT) Security," *International Conference on Computing for Sustainable Global Development (INDIACom)*, 2016.
6. Kim Thuat Nguyen, Maryline Laurent, Nouha Oualha, "Survey on secure communication protocols for the Internet of Things," Ad Hoc Networks, 2015.
7. Srinivasan Nagaraj, G.S.V.P. Raju, V. Srinadth. "Data Encryption and Authetication Using Public Key Approach," *International Conference on Intelligent Computing, Communication & Convergence*, 2015.
8. A. Islam, S. Y. Shin, BUS a blockchain-enabled data acquisition scheme with the assistance of UAV Swarm in Internet of Things. *IEEE Access* 7, 103231–103249, 2019.
9. Pawani Porambage, Corinna Schmitt, Pardeep Kumar, Andrei Gurtov, Mika Ylianttila. "Two-phase Authentication Protocol for Wireless Sensor Networks in Distributed IoT Applications," *IEEE WCNC'14 Track 3 (Mobile and Wireless Networks)*, 2014.
10. Zhi-Kai Zhang, Michael Cheng Yi Cho, Shiuhpyng Shieh. Emerging Security Threats and Countermeasures in IoT. National Chiao Tung University Hsinchu, Taiwan.
11. M. Young, *The Technical Writer's Handbook*. Mill Valley, CA: University Science, 1989.
12. A. Islam, S. Y. Shin, BUAV A blockchain based secure UAV-assisted data acquisition scheme in Internet of Things. *Journal of Communications and Networks* 21(5), 491–502, 2019.
13. Glederson Lessa dos Santos, Vinıcius Tavares Guimaraes, Guilherme da Cunha Rodrigues, Lisandro Zambenedetti Granville, Liane Margarida Rockenbach Tarouco. "A DTLS-based Security Architecture for the Internet," *20th IEEE Symposium on Computers and Communication (ISCC)*, 2015.
14. Thomas Kothmayr, Corinna Schmitt, Wen Hub, Michael Brünig, Georg Carle. DTLS based security and two-way authentication for the Internet of Things Ad Hoc Networks, 2013.
15. Corinna Schmitt, Thomas Kothmayr, Wen Hu, Burkhard Stiller. Two-Way Authentication for the Internet-of-Things. Ministry of Education and Research: The SODA Project under Grant Agreement No. 01IS09040A. and the AutHoNe Project under Grant Agreement No. 01BN070[25], 2012.
16. Kumar Priyan Malarvizhi, Usha Devi Gandhi. Enhanced DTLS with CoAP-based authentication scheme for the Internet of Things in healthcare application." *The Journal of Supercomputing*, 1–21, 2020.
17. Rolf H. Weber. Internet of Things – New security and privacy challenges. University of Zurich, Zurich, Switzerland, 2010.
18. Vijay Sivaraman, Hassan Habibi Gharakheili, Arun Vishwanath, Roksana Boreli, Olivier Mehani. "Network-Level Security and Privacy Control for Smart-Home IoT Devices," *Eight International Workshop on Selected Topics in Mobile and Wireless Computing*, 2015.

19. N.M. Mosharaf, Kabir Chowdhury, Raouf Boutaba. A survey of network virtualization. *Computer Networks* 54, 862–876, 2010.
20. V.C. Gungor, F.C. Lambert. A survey on communication networks for electric system automation. *Computer Networks* 50, 877–897, 2006.
21. Z. Liu, H. Seo. IoT-NUMS: Evaluating NUMS elliptic curve cryptography for IoT Platforms. *IEEE Transactions on Information Forensics and Security* 14(3), 720–729, 2019.
22. Rounak Sinha, Hemant Kumar Srivastava, Sumita Gupta. Performance based comparison study of RSA and elliptic curve cryptography. *International Journal of Scientific & Engineering Research* 4(5), 720–725, 2013.

Chapter 9

An efficient packet reachability-based trust management scheme in wireless sensor networks

Amit Kumar Gautam and Rakesh Kumar
Madan Mohan Malaviya University of Technology

CONTENTS

9.1 INTRODUCTION

Recent few years has witnessed an unpredictable growth in wireless communication technology. The wireless networks are categorized into two types such as infrastructure-based network and infrastructure less network. A wireless sensors network (WSN) is a type of infrastructure less network and group of small devices, known as sensor nodes. These nodes cooperate with other nodes to gather information from the environment. These tiny sensor nodes are composed of different modules viz; sensing module used for monitoring the environment, processing module for performing data processing, a communication module for transmitting data between sensor nodes, and power supply are used for energy. The architecture of a sensor node is depicted in Figure 9.1. A WSN is a resource constraint network. The network consists of nodes that are low cost, low processing power, limited energy, and limited storage capacity [1].

WSNs have gained significant amount of attraction from researchers in the field of academics as well as industrial communities. A WSN supports

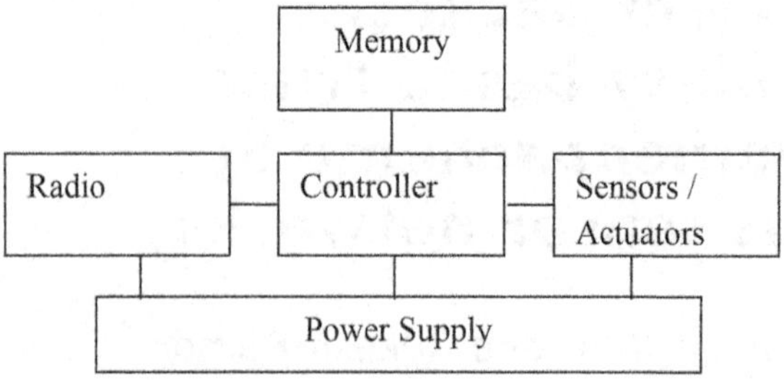

Figure 9.1 Architecture of sensor.

Table 9.1 WSN issues and challenges

Node deployment	Node deployment depends on a different type of application where the sensors are arranged in deterministic or randomized
Node heterogeneity	There are different types of sensors used in different applications
Energy	The sensor node has a limitation of energy due to limited battery sources
Scalability	Ability to work in small as well as large numbers of sensor networks
Mobility	WSN is dynamic due to the movement of sensors. Node movement causes frequent path breaks
Data delivery	WSN protocols have been affected by time, event, and data-driven reporting methods
Fault tolerance	WSN calibrate transmission powers on the link, if any node fails
Transmission media	Generally, the bandwidth used in the sensor for transmitting the data is (1–100 kb/s)
Converge-cast	Combination of data from different sources and collecting information "upwards" from the spanning tree after a broadcast

mobile communication requirements. For example, in industrial electronics, a sensor can be used for collecting and observing the data [2]. WSNs also have applications in many other fields such as health monitoring, intelligent building,

transportation, environmental surveillance, on-field sensing, and processing of signals and data aggregation. Table 9.1 shows the various issues and challenges in WSNs. The security in the sensor network is highly important not only for transferring the reliable data from sensor devices to distant sink but also for maintaining the network availability. Security is needed to ensure that the information which is sent from a sensor node is exactly the same which is received by the destination. By using different

security approaches in the WSN, we may ensure that the sensor node data remain confidential when some attack occurs during the transmission.

Thus, the data service is available in the network even the network is under attack. Thus, providing security in WSNs, we can get up-to-date data from the sensing field. WSN security needs data confidentiality, authentication, data integrity, availability, robustness, access control, and key management schemes. The solutions of security in WSNs include scalable, strong, lightweight, efficient, and effective key management and key distribution mechanisms and authentication. A WSN uses a single shared key which is insecure because an attacker can easily acquire the secret key. So, WSNs can establish the environment to distribute the key by using key management schemes [12].

WSN is vulnerable to various threats like internal threats and external threats. This threat causes the power exhaustion, network failure, information theft, intruder attacks, and many more. The solutions of security in the WSN include simple, robust, strong, low complexity, and efficient trust management. The WSN uses direct and indirect trust to secure themselves. To provide safe communication, the node should mitigate the malicious nodes with the help of the trust and reputation of the target node. Trust and reputation management should secure WSN applications and must provide scalability, authenticity, integrity confidentiality, and flexibility [13]. Several trust management models are proposed in the WSN. The nodes must calculate the trust value of other nodes. The trust value is updated dynamically and has more weight on recent transactions. The calculation of trust solely depends upon the packet transaction among nodes. The cluster head plays as a recommendation manager which helps to calculate the indirect trust of any node. In this paper, we propose a packet reachability-based trust management scheme that applies to WSN communication which has constraints like storage, power, and computation. The main contributions of the paper are as follows:

- Study of salient features of existing trust management schemes.
- Generate and distribute trust values with the help of packet reachability of each node.
- Time lapses function is used to calculate direct trust while recommendation-based feedback is used to calculate indirect trust.
- Simulation has been carried out to demonstrate the effectiveness of the proposed work.

The remaining part of this paper is arranged as follows. Section 9.2 presents the background of this paper while Section 9.3 gives the various works in the related area of the trust calculation. The proposed method of reachability-based trust calculation is discussed in Section 9.4. Section 9.5 describes the performance evaluation. Finally, the concluding remarks with future directions are presented in Section 9.6.

9.2 BACKGROUND

9.2.1 Trust

Many definitions of trust are proposed by the authors. Trust is defined as the level of belief that is developed by past interaction and behavior between source and destination nodes. Trust has made an impact on future route selection and communication [13,14]. Direct trust is established between the source node and neighbor nodes. The direct communication between nodes helps to evaluate the direct trust. It has more impact than indirect and recommendation-based trust. Indirect trust establishes when the source node cannot directly connect to the target node and observes the behaviors of the target node through other nodes. This is the combination of feedback and recommendations of other nodes. Recommended trust is part of indirect trust where the neighbors of the target node give feedback about that node. It is based on the trust record of direct neighbors about the target node based on their experiences with the target node [15].

9.2.2 Security attacks

Based on the behavior of the adversary node, some security attacks can be illustrated as follows [12,15,16]:

- *Black hole*: In black hole attack, a node falsifies the route of network and attracts all the packets and routing information towards itself.
- *Selective forwarding*: This type of attack consists a malicious node that behaves like a router and drops some packets and may deny forwarding that packets or messages.
- *Bad mouthing attack*: In this type of attack, the malicious node gives the wrong information about the neighbor node.
- *Denial of service (DoS) attacks*: In DoS attack, the malicious node injects bad information to mislead the network. Here, the bad node provides the wrong reputation and feedback about other nodes.
- *Sybil attack*: In this, the malicious node has many IDs and behaves as like many nodes.
- *Wormhole attack*: Attackers strategically placed malicious nodes at different ends of a network. They can create a tunnel and receive messages. After receiving, it forwards to other nodes.
- *Identity replication attack*: In this attack, the attacker can make a copy of the existing node and place itself to different parts of a network. It can create multiple copies of the same node at different places. So, when this type of attack happens, then, no way to identify that the network is compromised.
- *Collusion attack*: In this type of attack, more than one malicious node gives false information and feedback about good nodes.

9.2.3 Motivation

The motivation behind the proposition of a new trust management model is that the previously proposed trust management methods are costly and have higher communication complexity and storage overhead. Our proposed model is a packet reachability-based approach which is lightweight and takes less computation complexity. It emphasizes more than recent communication and successful packet delivery rather than old communication details. One of the main objectives is to devise a model to be generic, independent of the semantics assigned to trust. The proposed scheme uniformly supports the nodes with high diverse capabilities and distributes the computation cost of trust computation revocation [14,17–20]. There are various security challenges in WSNs. Due to the following characteristics, the traditional method of security cannot efficiently be applied to the WSN.

- The WSN should be economically feasible as sensor devices have limited storage, small computation, low energy, and communication capabilities.
- The WSNs have been deployed in a remote area which presents extra risks of physical attacks.
- Due to the broadcast nature of WSNs, normally it increases challenges to security.

9.3 RELATED WORK

Many researchers have proposed in the evaluation of trust in the field of P2P network, cloud computing, and WSNs. The previous trust management models [21,22] mainly focus on wired networks like the P2P network, and conventional trust management models [23,24] developed for wireless ad-hoc networks. So, these schemes are not suitable for WSNs because they are not efficient in terms of memory and power. Due to memory and power constraints, some trust management schemes, e.g., PLUS [9] and ATRM [25], are specially designed for WSNs. Some recently proposed trust management [26,27] schemes based on fuzziness and soft computing approaches provide security in sensor networks.

Gadde et al. [3] solved issues like low security, high packet loss rates, and maximized latency. The author proposed a reliable data delivery-based trust calculation scheme. In this trust management system, each node has a trust value which is increasing or decreasing based on the successful delivery of the packets. Depending on these trust values, nodes choose the trusted path to transfer data. It reduces the data size, energy consumption, and storage overhead by compressing the packet before being sent. Therefore, it uses a cryptographic approach to gain security. A game theory-based trust management scheme is proposed by Yang et al. [4] who observe the dynamic

behavior of nodes by a dynamic collection of evidences. In their proposed method, the trust mechanism integrated into the cluster-based routing protocol. It gives high network lifetime and security.

A group-based trust management scheme (GTMS) is proposed by Shaikh et al. [5], and it is designed to defend against a black hole attack. To compute trust, GTMS is adopted to collect recommendations from all the nodes and calculates total trust value. Here, the total trust is a combination of direct and indirect monitoring of nodes. Each cluster head has the collection of recommendations about any node inside their cluster. The sink node collects recommendations from all cluster heads and chooses the best-trusted route. GTMS takes less memory, and trust is represented as an unsigned integer between 0 to 100. The drawback of GTMS is that it requires high-energy Cluster Heads (CHs) to directly communicate with the sink.

Zahariadis et al. [6] proposed a trust management scheme to prevent bad mouthing sybil attack and conflicting behavior. The trust value is represented in the range between 0 and 1. A lightweight and dependable trust system [7] is designed for hierarchical WSNs to thwart black hole and bad mouthing attacks. The trust was computed based on direct and indirect observations. A centralized trust management scheme was used in intra cluster and inter cluster level. The trust value was assigned in the range of 0 to 10. All the above-mentioned schemes use a promiscuous mode of operation for direct observation. The malicious nodes were identified only based on the past experience of a node. Jiang et al. [8] proposed an efficient distributed trust model (EDTM) for WSNs. EDTM uses the number of packets for trust calculation. Direct trust and recommendation-based trust are calculated using the number of packets transferred between nodes. This is an efficient mechanism to calculate the trust in a distributed environment of WSNs. Yao et al. [9] presented a comparative study of various attacks and their countermeasures associated with trust schemes. However, they did not present a secure routing protocol according to the analysis result.

Josang et al. [10] proposed a model based on Bayesian distribution and follow binomial and multinomial rating in the dynamic environment. In binomial case, it uses beta distribution function while in multinomial, it uses Dirichlet probability density function. Xuan et al. proposed a method where two types of trust tables are used: direct trust table and recommended trust table. In this model, when a source node is looking for the trust value of the target node, then it goes to check direct trust first, if it is available, then direct trust is the final trust. If direct trust is not available, then indirect trust is the final trust. Khan et al. [11] proposed a trust estimation method for a bulky sensor network. WSN retains grouping to recover cooperativeness, honesty, and security by detecting faulty nodes and reducing resource consumption. The author believes that they provide unique IDs for sensors to protect sensors from external attacks. It follows the distributed and centralized sensor network for inter-cluster and intra cluster networks to find the trust between nodes. A timing gap analysis is

used to find out a successful or unsuccessful transaction. In this work, the author has not considered weight, frequency of fault trust, on–off attack, DoS, and collusion attacks. A novel and secure trust management approach [28] is proposed for scalable WSNs. In this method, the authors proposed a trust management method by using time lapses function. The time lapses function is used to give more weight for recent communication than older communication. The direct trust is evaluated by lightweight mathematical functions where it takes less resources. The trust has been updated dynamically and has more weight on recent transactions. The calculation of trust has been fully depended upon the packet transactions among nodes. They assumed that the environmental factor is constant during the transaction. Table 9.2 depicts the salient features of existing trust management schemes in the literature.

9.4 PROPOSED TRUST MODEL

In the proposed model, every sensor node is responsible to publish various services like trust and repudiation, quality of service like residual energy, storage cost, reliability, efficiency, etc. The trust is either direct or indirect. Direct trust has been calculated by directly neighbor nodes which depends upon the successful transmission of a packet. The indirect trust uses feedback and other nodes opinions which are not direct neighbors of that node. In a hierarchical sensor network, the cluster head has the responsibility to store the trust value of other nodes in the cluster [5]. The total trust is calculated by adding the direct trust and recommendation-based trust, i.e., indirect trust. Based on the total trust, the source node chooses the secure path and prevents against internal threats. Figure 9.2 depicts the proposed trust model.

Direct trust is established between the source node and neighbor nodes. The direct communication between nodes helps to calculate the direct trust. It has more impact than indirect and recommendation-based trust. In this model, we have assumed that the recent conversations have more weight than past conversations. It means that the oldest conversation has negligible weight while the most recent conversation has the largest weight [8]. First, we find the neighbors between source and destination node. After that, we calculate the direct trust value of each neighbor node. Suppose, A is the source node and B is one of its neighbor nodes. If the number of transactions between nodes A and B is p, and s is the number of packets successfully received by node B, then, p-s number of packets shall be dropped.

Then, a successful transaction between A and B is calculated as follows.

$$T(x) = \frac{s}{q+s} \text{ where } p = q + s \tag{9.1}$$

Table 9.2 Summary of salient features of existing trust management schemes

Trust scheme	*Methodology*	*Structure*	*Security agent*	*Advantages*	*Limitations*
Gadde et al. [3]	Reliability-based trust management	Distributed	Reliability	Reduces the data size, energy consumption, and storage overhead	Depends upon reliability of packet
Yang et al. [4].	Observes the dynamic behavior of nodes using a dynamic collection of evidences	Cluster-based and centralized	Dynamic behavior	High network lifetime	Trust estimation and convergence time is high
Shaikh et al. [5]	Multilevel trust calculation at three levels such as sensor, cluster head, and sink node level	Cluster-based, centralized, and distributed	Time window	Less computation and storage overhead	Not effective against on–off attack and not fit for real-time applications
Zahariadis et al. [6]	Geographical data-based routing	Distributed	Geographical data	Defends against bad mouthing Sybil attack, and conflicting behavior	Not proven the system robustness
Li et al. [7]	Weighted value for transaction and recommendation	Centralized and distributed	History of transactions, recommendation	Complexity is minimum	A weighted coefficient is vulnerable for attacks
Jiang et al. [8]	Trust is calculated using the number of packets transferred between nodes	Distributed	No. of packets successfully transferred	Energy-efficient in a distributed environment	Not accurate in cluster or centralized environment
Yao et al. [9]	Direct and indirect trust calculation locally	Distributed	Recommendation and personal reference	Efficient in detection of faulty nodes and also provides real-time security	Trust estimation and convergence time is high
Josang et al. [10]	Bayesian distribution and follow binomial and multinomial rating in a dynamic environment	Distributed	Bayesian distribution and Dirichlet probability density function	Less computation and storage overhead	Misleading by several attacks like node compromised attack
Khan et al. [11]	Calculates trust node to node, cluster head to cluster head, and node to cluster head	Distributed approach and centralized	Direct and feedback	Fast, energy-efficient, and low computation overhead	Vulnerability against DoS attack and several security attacks

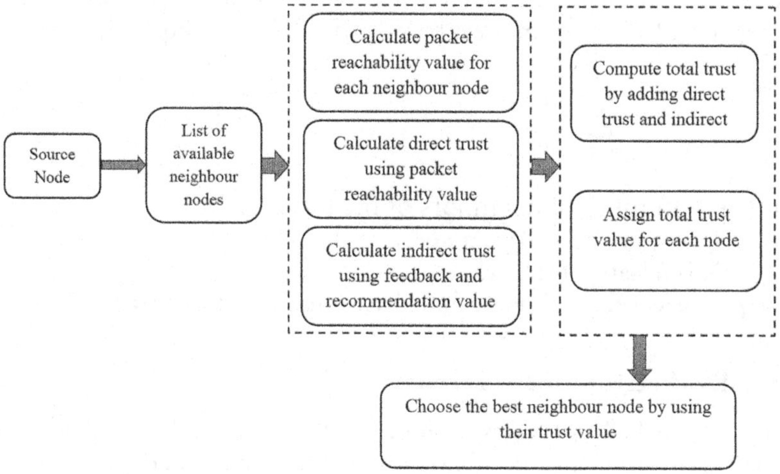

Figure 9.2 Proposed trust model.

So, the direct trust value of between node A to node B is calculated as follows.

$$\mathrm{TR}_{\mathrm{dir}}^{n}(A, B) = u(T(x)) + (1-u)\,\mathrm{TR}_{\mathrm{dir}}^{n-1}(A, B) \quad (9.2)$$

where $\mathrm{TR}_{\mathrm{dir}}^{n}(A, B)$ is the direct trust of node A on node B at the nth transaction between node A and node B, u is the time lapses function [22] and is defined as the dynamic factor which changes according to the following Eq. (9.3).

$$u = f(x) = \begin{cases} 1 - \left(\dfrac{t_{n-1} - t_1}{t_n - t_1}\right)^2, & t_n > t_1 \\ 1, & \text{otherwise} \end{cases} \quad (9.3)$$

The value of u is higher if the transaction is more recent one while it is lesser if the transaction had been established long ago.

9.4.1 Recommendation (feedback)-based trust

When the source node is not connected directly with the target node, then, the trust value of the target node is calculated indirectly. These indirect trusts are evaluated from the feedback of target's neighbor nodes. The steps to calculate indirect trust value are as follows:

Step 1: Source node A requests the neighbors of target node B for feedback
Step 2: Receives the response feedback from the neighbors of target node B

Step 3: Filter the false nodes feedback by removing outlier value of trust
Step 4: Calculate the indirect trust by the following Eq. (9.4).

$$\mathrm{IDT}(A, B) = \sum_{i=1}^{n} \frac{P}{P+N} \tag{9.4}$$

where n = number of neighbors of the target node,
P is positive feedback and
N is negative feedback.
Step 5: Select the legitimate node based on the indirect trust.

9.4.2 Evaluation of total trust

In our approach, $T_{A,B}$ denotes the total trust between node A and node B at a particular time t. It is the addition of direct trust and indirect trust. The direct trust is $\mathrm{TR}_{\mathrm{dir}}(A, B)$, and the indirect trust $\mathrm{IDT}(A, B)$ is calculated for networks in each round of communications using Eqs. (9.2) and (9.4) as mentioned above. Total trust is calculated using the following Eq. (9.5).

$$T_{A,B} = c\ \mathrm{TR}_{\mathrm{dir}}(A, B) + (1-c)\ \mathrm{IDT}(A, B) \tag{9.5}$$

where ‘c’ is a constant which varies from application to application and $c \in (0,1)$. It balances the value of direct trust and indirect trust. If $c = 1$, $T_{A,B}$ is equal to direct trust. If $c = 0$, then $T_{A,B}$ is equal to indirect trust.

9.5 PERFORMANCE EVALUATION

The proposed method is simulated using the ns2 simulator for packet reachability. Based on these values, we separately calculated the trust value. Table 9.3 shows the simulation parameters used in the proposed approach.

Table 9.3 Simulation parameters

Terms	*Value*
Simulator	ns^{-2}
Number of nodes	10–90
Simulation area	1,000 × 1,000 m
Transmission range	250 m
Routing protocol	SEP protocol
Simulation time	200 s
Packet size	1024 bytes
Buffer size (queue length)	50 Pkts

The results of the proposed method have been evaluated by using a packet reachability metric to prove its robustness and efficiency. An increase and decrease of trust value depends on the successful transmission of a packet from a source node to the target node. Based on the trust value, we removed the nodes which give consistently the lowest value of trust. Our trust model is more secure and efficient compared with the traditional model. We have taken the scenario in which the packet transfer efficiency continuously increases, decreases, or both. Consider the various packets reaching the probability, as illustrated in Figure 9.3. Data set A has continuously increasing packet transfer probability. Data set B has continuously decreasing packet transfer probability. Data set C has the data where the probability of the packet reached is uncertain.

In Figure 9.3, our trust model successfully shows the increase and decrease of trust values of nodes according to time lapses function [22] and

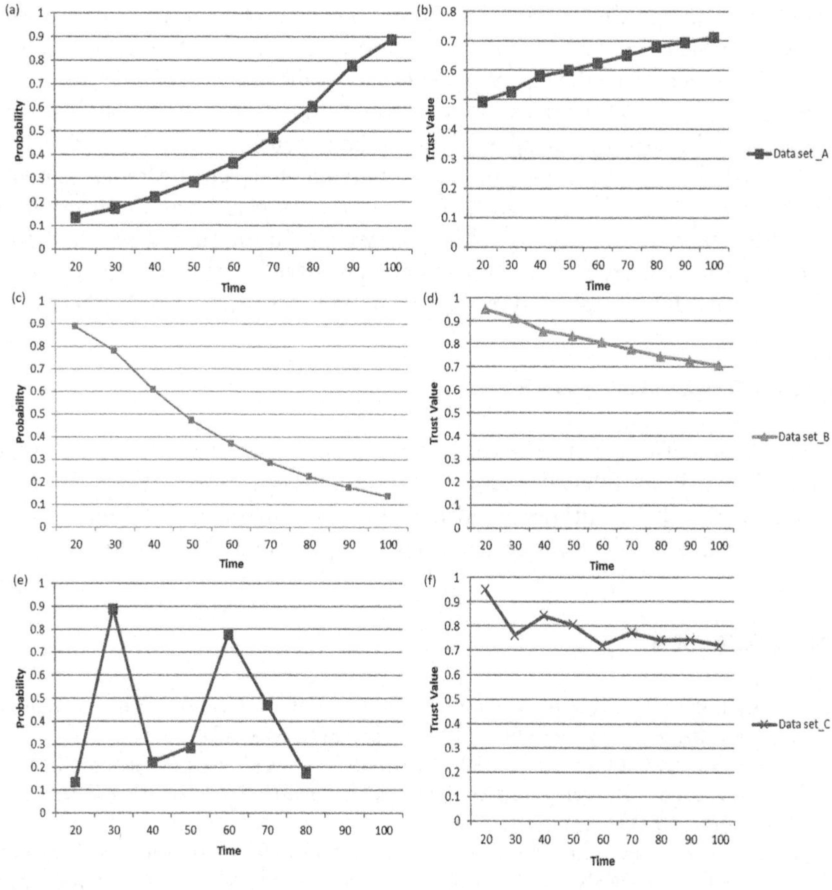

Figure 9.3 Packet reachability probability and corresponding total trust value.

packet reachability. Figure 9.2a represents the packet reachability probability which is continuously increasing, and its corresponding trust value is depicted in Figure 9.3b. Similarly, Figure 9.2c represents the packet reachability probability which is continuously decreasing, and its corresponding trust value is depicted in Figure 9.3d. Figure 9.3e represents the data which are having random behavior, and its corresponding trust value is shown in Figure 9.3f.

9.6 CONCLUSIONS AND FUTURE WORK

WSNs are always under the threat of various malicious attacks. There are many attacks like bad mouthing attack, on–off attack, self-promotion attack, Sybil attack, etc. which occur due to the false trust value. In this paper, we have proposed a simple and secure trust model to secure the network. The proposed model is based on the trust value which takes into consideration the direct trust and indirect trust. The total trust is calculated with the help of packet reachability between nodes. It uses time lapses function which gives the recent packet reachability more emphasis than the past conversations. Our trust model is robust and is also secure against collusion attack, self-promoting attack, and bad mouthing attacks.

Nevertheless, the same proposed trust management scheme for a session in WSNs can be extended efficiently for a multi-session scenario in the domain of WSNs or in the Internet of Things networks. In the future, we shall add Service Oriented Architecture and some new attributes like reliability, scalability, and fault tolerance in our model to enhance security in WSNs.

ACKNOWLEDGMENTS

This paper is funded by the University Grant Commission (UGC), India under Junior Research Fellowship (UGC NET-JRF) vide letter no. 3331/(SC) (NET-JUNE2015). Authors are also thankful to Director, IMS Engineering College, Ghaziabad for his constant encouragement and motivation.

REFERENCES

1. Akyildiz IF, Su W, Sankarasubramaniam Y, Cayirci E (2002) Wireless sensor networks: a survey. *Computer Networks* 15(4), 393–422.
2. Agarwal R, Gautam AK (2012) A probability based energy efficient clustering protocol in wireless sensor network. *International Journal of Advanced Computer Technology* 1(1), 4–9.
3. Gadde D, Chaudhari MS (2018) Reliable data delivery on the basis of trust evaluation in WSN. In *Information and Communication Technology for Sustainable Development*. Springer, Singapore, 51–59.

4. Yang L, Lu Y, Liu S, Guo T, Liang Z (2018) A dynamic behavior monitoring game-based trust evaluation scheme for clustering in wireless sensor networks. *IEEE Access* 6, 71404–12.
5. Shaikh RA, Jameel H, d'Auriol BJ, Lee H, Lee S, Song YJ (2009) Group-based trust management scheme for clustered wireless sensor networks. *IEEE Transactions on Parallel and Distributed Systems* 20(11), 1698–712.
6. Zahariadis T, Trakadas P, Leligou HC, Maniatis S, Karkazis P (2013) A novel trust-aware geographical routing scheme for wireless sensor networks. *Wireless Personal Communications* 69(2), 805–26.
7. Li X, Zhou F, Du J (2013) LDTS: A lightweight and dependable trust system for clustered wireless sensor networks. *IEEE Transactions on Information Forensics and Security* 8(6), 924–35.
8. Jiang J, Han G, Wang F, Shu L, Guizani M (2015) An efficient distributed trust model for wireless sensor networks. *IEEE Transactions on Parallel & Distributed Systems* 1(1), 1–1.
9. Yao Z, Kim D, Doh Y (2006) PLUS: Parameterized and localized trust management scheme for sensor networks security, In 2006 *IEEE International Conference on Mobile Ad Hoc and Sensor Systems*, pp. 437–446. IEEE.
10. Jøsang A, Ismail R, Boyd C (2007) A survey of trust and reputation systems for online service provision. *Decision Support Systems* 43(2), 618–44.
11. Khan T, Singh K, Abdel-Basset M, Long HV, Singh SP, Manjul M (2019) A novel and comprehensive e trust estimation clustering based approach for large scale wireless sensor networks. *IEEE Access* 7, 58221–40.
12. Pathan AS, Lee HW, Hong CS (2006) Security in wireless sensor networks: issues and challenges. In *Advanced Communication Technology, 2006. ICACT 2006. The 8th International Conference 2006*. Feb 20, 2, IEEE, 6–13.
13. Han G, Jiang J, Shu L, Niu J, Chao HC (2014) Management and applications of trust in wireless sensor networks: a survey. *Journal of Computer and System Sciences* 80(3), 602–17.
14. Rodrigues P, John J. (2020) Joint trust: an approach for trust-aware routing in WSN. *Wireless Networks*, 26, 1–6.
15. Selvi M, Thangaramya K, Ganapathy S, Kulothungan K, Nehemiah HK, Kannan A. (2019) An energy aware trust based secure routing algorithm for effective communication in wireless sensor networks. *Wireless Personal Communications* 105(4), 1475–90.
16. Kumar V, Kumar R (2015) An adaptive approach for detection of blackhole attack in mobile Ad hoc network. *Procedia Computer Science* 48, 472–79.
17. Reddy, VB., Negi A and Venkataraman S (2018, October). Trust computation model using hysteresis curve for wireless sensor networks. In *2018 IEEE SENSORS* pp. 1–4. IEEE.
18. Singh O, and Rishiwal V. (2020). Scalable energy efficient routing mechanism prolonging network lifetime in wireless sensor networks. *International Journal of Systems, Control and Communications*, 11(2), 161–177.
19. Chahal RK, Kumar N, Batra S. (2020) Trust management in social Internet of Things: a taxonomy, open issues, and challenges. *Computer Communications* 150, 13–46.
20. Srikanth N, Prasad MS. A compressive family based efficient trust routing protocol (C-FETRP) for maximizing the lifetime of WSN. *Data Communication and Networks* 2020. Springer, Singapore, 69–80.

21. Gupta M, Judge P, Ammar M (2003) A reputation system for peer-to-peer networks. In *Proceedings of the 13th International Workshop on Network and Operating Systems Support for Digital Audio and Video*. Association for Computing Machinery, New York, NY, United States, 144–52.
22. Kamvar SD, Schlosser MT, Garcia-Molina H (2003) The eigentrust algorithm for reputation management in p2p networks. In *Proceedings of the 12th International Conference on World Wide Web*. May 20, Association for Computing Machinery, New York, NY, United States, 640–51.
23. Liu Z, Joy AW, Thompson RA (2004) A dynamic trust model for mobile ad hoc networks. In Distributed Computing Systems, 2004. FTDCS 2004. In *Proceedings of 10th IEEE International Workshop on Future Trends of 2004*. May 26, IEEE, Los Alamitos, CA, 80–85.
24. Pirzada AA, McDonald C (2004) Establishing trust in pure ad-hoc networks. In *Proceedings of the 27th Australasian conference on Computer Science*. Australian Computer Society, Inc., Australian Computer Society, Darlinghurst, NSW, Australia, 47–54.
25. Boukerch A, Xu L, El-Khatib K (2007) Trust-based security for wireless ad hoc and sensor networks. *Computer Communications* 30(11–12), 2413–27.
26. Ya W, Meng-Ran Z, Lei N, Jia Z. (2020) Trust analysis of WSN nodes based on fuzzy theory. *International Journal of Computers and Applications* 42(1): 52–6.
27. Das R, Dash D, Sarkar MK. (2020) HTMS: fuzzy based hierarchical trust management scheme in WSN. *Wireless Personal Communications*, 112, 1–34.
28. Gautam AK, Kumar R (2018) A robust trust model for wireless sensor networks. In *Proceedings of 5th IEEE Uttar Pradesh Section International Conference Electrical, Electronics Computer Engineering (UPCON)*., pp. 1–5, IEEE.

Chapter 10

Spatial domain steganographic method detection using kernel extreme learning machine

Shaveta Chutani
IK Gujral Punjab Technical University

Anjali Goyal
GNIMT

CONTENTS

10.1 INTRODUCTION

Digital images have become the essence of all digital communication over internet. This leads to ever-increasing chances of images being used by unscrupulous elements for illegitimate transactions as well. Steganography is such a technique of embedding the covert message into cover object which appears innocent (Petitcolas et al. 1999). The main goal of steganography is to hide secret information without arousing suspicion. Steganalysis is the counter technique to defeat steganography (Böhme 2010). It aims to discover the secret hidden message from the cover image. Steganalysis can be broadly categorized as active and passive (Chandramouli 2003). While the passive steganalysis deals with determining the presence of secret information and identifying the steganographic algorithm used to embed; active steganalysis further involves estimating the length, finding message bearing pixel locations, determining the stego key used in embedding, and ultimately extracting the stego message itself (Trivedi and Chandramouli 2003).

Passive steganalysis sets the background for active steganalysis. It provides the important information requirements for further stages in active steganalysis so that the message extraction is more accurate and computationally less expensive. Hence, it plays a crucial role in achieving the ultimate objective of message extraction. While the first goal of passive steganalysis can be achieved through two-class classification separating innocent cover images from stego images carrying the hidden content, the second goal can be seen as an extension of the first one. This objective usually employs multi-classifier to distinguish not only cover and stego images but also the type of steganographic algorithm used to create stego images. Both of these objectives exploit different feature sets of innocent cover and stego images. A machine learning algorithm learns these feature set patterns and classifies these images on the basis of differences in these.

In this paper, we propose the use of kernel extreme learning machine (KELM) (Guang-Bin Huang et al. 2012) and subtractive pixel adjacency matrix (SPAM) (Pevný et al. 2010) features to determine the spatial domain steganographic algorithm applied to embed the secret message. The paper is structured as follows. Section 10.2 discusses the related works as available in the steganalysis literature. The motivation to use KELM as a learning tool is also presented. The basic concepts of ELM and KELM are presented in Section 10.3. Details about the SPAM features are also presented in the same section. Section 10.4 illustrates the proposed methodology while Section 10.5 presents the details of the experiments carried and their results. The paper is concluded in Section 10.6.

10.2 RELATED WORK

One of the earliest multi-class steganalyzers was proposed by Pevný and Fridrich (2005). The authors use calibrated discrete cosine transform (DCT) features of single compressed JPEG images and use support vector machine (SVM) with Gaussian kernel for multi-class classification. Later, Pevný and Fridrich (2007) presented a multi-class steganalyzer for JPEG images exploiting merged Markov and extended DCT features coupled with soft margin C-SVM as a classifier. Dong et al. (2009) suggested using image run-length analysis features and Radial Basis Function (RBF) kernel SVM for hierarchical multi-class classification, classifying cover and stego in the first stage and identifying embedding algorithm in the next stage. Zhu et al. (2015) used CC-JRM feature set with a linear SVM in two different ensemble strategies. In the first strategy, the authors use multiple multi-class linear SVM classifiers and use majority voting for classification; whereas in the second, they use an ensemble of binary linear SVM classifiers in place of ordinary binary classifiers of the conventional multi-classifier. Rodriguez and Peterson (2008) fused classification results from four different classifiers using boosting. The three of these systems

are multi-classifiers trained on different feature sets viz. wavelet features, DCT features, and combined DCT features. They use StegDetect as the fourth classifier. Veena et al. (2019) proposed a new set of hybrid features, local residual pattern and local distance pattern for steganographic method detection. The authors optimized these features further using a Greedy Randomized Adaptive Search – Binary Grey Wolf Optimization technique and tested these optimized features on different spatial embedding algorithms using different classifiers.

Cho et al. (2010) suggested a block-based approach for steganalysis. In this, the image in the training set is decomposed into homogenous blocks and merged Markov and DCT features are extracted from each of these blocks. These blocks are randomly sampled and classified into multiple classes based on the tree structured vector quantization method. The image is classified based on the fusion of weighted results of all blocks. Lubenko and Ker (2011) used logistic regression as the machine learning tool using SPAM features for spatial domain embedding algorithms. The authors present results of both, i.e., linear and kernelized variants logistic regression algorithms for multi-class classification. Xu et al. (2018) used convolution Neural Network (CNN) for steganographic algorithm detection. The authors generate a match image from the stego and extract deep features by embedding it with different algorithms. The features are then compared with the stego image for recognizing the embedding method.

We can see that previous works discussed above can be aligned in two broad directions: first, in which researchers experimented using different image feature sets and their combinations with SVM as the learning algorithm and second, where researchers deviated from the norm of using SVM for classification and rather used other machine learning tools and frameworks. Deriving motivation from this, for the present paper, we use KELM for this multi-class classification objective. Extreme learning machine (ELM) is a single-layer feedforward network proposed by Huang et al. (2006). It offers great scalability, generalizability, and performs impartially for binary classification, multi-class classification, and regression problems.

10.3 BACKGROUND CONCEPTS

The steganalysis space hitherto has been dominated by SVMs as the machine learning tool for classification problems. Interestingly, SVMs are inherently binary classifiers but different researchers used these for multi-class classification by constructing different architectures such as using multiple SVMs for individual binary classification and then aggregating results for the overall multi-class classification tasks and using the kernel trick for nonlinear classification, etc. Another inherent limitation of the SVMs is their slow performance. ELMs are new machine learning tools which have been successfully applied to various classification and regression tasks. ELMs are

intrinsically fast and do not require much human intervention. The kernel ELMs even do away with the limitation of random input weights and bias.

The performance of any machine learning algorithm is equally dependent on the feature set used. For universal as opposed to targeted steganalysis, the characteristics of the feature set play an important role. SPAM feature set effectively captures the statistical relationships and inter-pixel dependencies of stego images. This feature set has been widely used for steganalysis of spatial domain embedding algorithms. In the following paragraphs, we present some background concepts related to ELM, kernel ELM, and SPAM feature set.

10.3.1 Extreme learning machine

ELM (Huang et al. 2006) is a neural network architecture with two stages for learning. The brief architecture of ELM can be summarized, as shown in Figure 10.1.

Assuming an image dataset consisting of N instances of images with d-dimensional feature set, the input space to a typical ELM is (X_i, T_i) where $X_i \in R^{N \times d}$ with target class, $T_i \in R^{N \times c}$ where c is the number of target classes.

The output function for a single-layer feed-forward ELM network with L hidden layer neurons and $g(x)$ activation function can be represented as in Eq. (10.1):

$$\sum_{j=1}^{L} \beta_j g_j(x_j) = \sum_{j=1}^{L} \beta_j g(w_j.x_i + b_j) = o_i \tag{10.1}$$

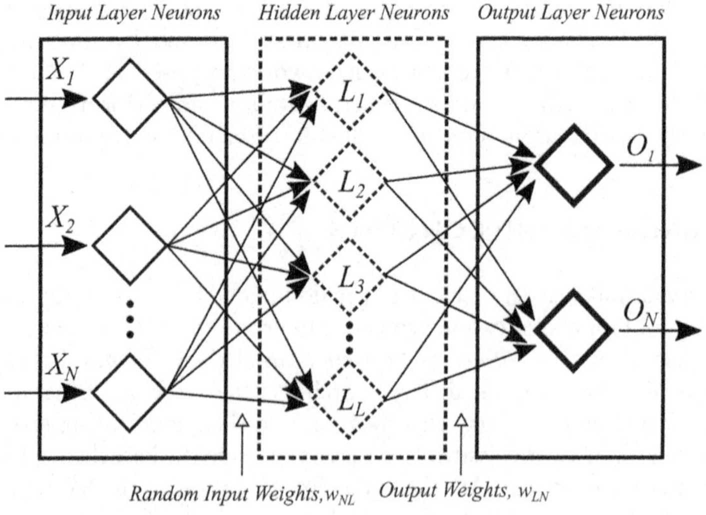

Figure 10.1 ELM architecture.

where $i = 1,\ldots, N$; $w_j = \left[w_{j1}, w_{j2}, \ldots, w_{jn}\right]^T$ is the weight vector connecting the input layer with jth hidden layer neuron; $\beta_j = \left[\beta_{j1}, \beta_{j2}, \ldots, \beta_{jn}\right]^T$ is the weight vector connecting the jth hidden layer neuron with the output layero_i; and b_j is the threshold of the jth hidden layer neuron. Input weights, w, and biases, b, are randomly initialized.

The standard ELM with L hidden layer neurons and $g(x)$ activation function can approximate N instances with zero error which can be written as in Eq. (10.2):

$$H\beta = T \tag{10.2}$$

Output weights, β, can be analytically determined by finding the unique smallest norm least-squares solution of the linear system (10.3):

$$\beta = H^{\dagger}Y \tag{10.3}$$

where $H^{\dagger}$is the Moore-Penrose generalized inverse of matrix H and $H^{\dagger} = H^T\left(HH^T\right)^{-1}$. Huang et al. (2006) proposed to add a positive value 1/C (where C is a user-defined and optimized hyper parameter) for the calculation of the output weights β such that

$$\beta = H^T\left(\frac{1}{C} + HH^T\right)^{-1} Y \tag{10.4}$$

10.3.2 Kernel ELM

Huang et al. (2012) proposed to substitute a kernel function for the inner product by applying the Mercer condition to ELM. It can be represented as in Eq. (10.5):

$$\Omega_{\text{ELM}} = \begin{bmatrix} k(X_i, X_i) \\ \vdots \\ k(X_i, X_i) \end{bmatrix} = HH^T;\ \Omega_{\text{ELM}_{i,j}} = h(X_i).h(X_i) = k(X_i, X_i) \tag{10.5}$$

where $k(X_i, X_i)$ is a kernel function. The output function can be written as in Eq. (10.6)

$$\begin{bmatrix} k(X_i, X_i) \\ \vdots \\ k(X_i, X_i) \end{bmatrix}^T .(\frac{1}{C} + \Omega_{\text{ELM}})^{-1}Y \tag{10.6}$$

Different kernel functions such as polynomial, RBF, wavelet function can be used with kernel-based ELM.

10.3.3 SPAM (subtractive pixel adjacency matrix) feature set

This feature set was proposed by Fridrich et al. mentioned in the study by Pevný et al. (2010). The local dependencies between differences of neighboring pixel values of an image are modeled as first-order and second-order Markov chains. The sample probability transition matrix then computed constitutes the feature space. These feature extraction steps are as follows:

1. Calculate the difference array D for all eight directions, i.e., two along each horizontal, vertical, major, and minor diagonal directions:

$$D_{\overline{i,j}} = I_{i,j} - I_{i,j+1} i \in \{1, \ldots, m\},\ j \in \{1, \ldots, n-1\} \tag{10.7}$$

2. Model the first-order and second-order SPAM features by the Markov process.

$$M_{\overline{x,y}} = \Pr\left(D_{\overline{i,j+1}} = x \mid D_{\overline{i,j}} = y\right) \tag{10.8}$$

$$M_{\overline{x,y,z}} = \Pr\left(D_{\overline{i,j+2}} = x \mid D_{\overline{i,j+1}} = y,\ \mid D_{\overline{i,j}} = z\right) \tag{10.9}$$

where x, y, $z \in \{-T, \ldots, T\}$.

3. Horizontal and vertical matrices and both diagonal matrices are separately averaged to get the final features, i.e., F^{1st} and F^{2nd}.

The authors showed that the value of T should be optimum to trade-off between classification accuracy and computational complexity. They further demonstrate that increasing T in the first order features does not necessarily improve performance of any steganalyzer but the second order features extracted by fixing $T = 3$ exhibit much improved performance despite increased dimensionality of 686.

10.4 PROPOSED METHODOLOGY

The methodology used in this paper is presented in Figure 10.2.

The process of steganographic algorithm identification involves the following steps:

a. *Feature set extraction*: The SPAM features of the image set are extracted.

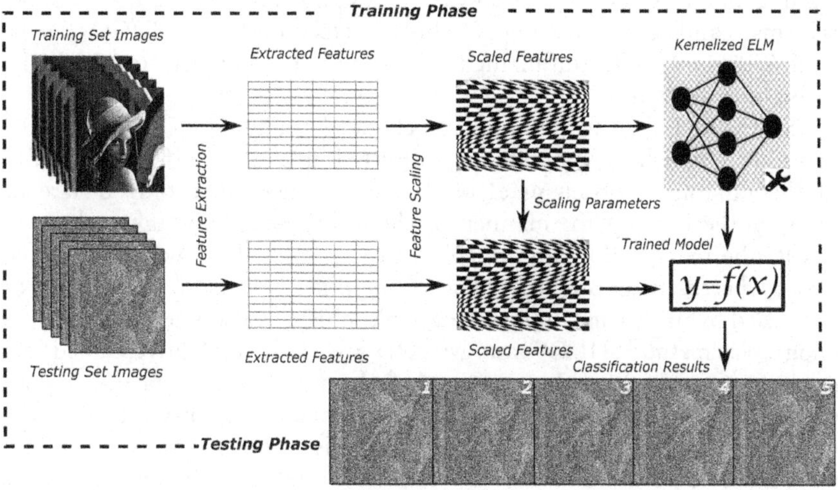

Figure 10.2 Multi-class classification using KELM.

b. *Feature pre-processing*: The extracted features are scaled in the range of 0 to +1.
c. *KELM classifier training*: The KELM is trained using the training set cover and stego images.
d. *Hyper parameter optimization*: Hyper parameters are searched using the grid search method and 5-fold cross-validation is used for hyper optimization. Here, Gaussian Radial Basis Function is used as the kernel function given by Eq. (10.10)

$$K(a,b) = \exp\left(-\gamma \left\|a-b\right\|\right)^2 \tag{10.10}$$

Two main parameters to be optimized are the penalty parameter C in Eq. (10.4) and the kernel bandwidth γ in Eq. (10.10). The parameter C determines the trade-off between model complexity and the fitting error minimization. The parameter γ defines the non-linear mapping from the input space to high-dimensional feature space.
e. *Test set classification*: Using the trained classifier's model, the unseen test images are classified as either cover or different stegos, thus identifying the embedding algorithm.

10.5 EXPERIMENTAL SETUP AND RESULTS

Experiments are carried out on BOSSBase v1.01 (BOSS Web page 2010) image dataset consisting of 10,000 images. The stego images are generated using four different spatial domain embedding algorithms viz. Least

Significant Bit Replacement (LSBR),[1] Least Significant Bit Matching (LSBM) (Harmsen and Pearlman 2003), Highly Undetectable Stego (HUGO) (Pevný et al. 2010a, b), and Minimizing the Power of Optimal Detector (MiPOD) (Sedighi et al. 2016).

LSBR and LSBM are among the classical spatial domain embedding methods in which the message bit is inserted in Least Significant Bit (LSB) of the randomly chosen pixel of the cover image. The only difference between these two is the manner in which this embedding takes place. In LSBR, the message bit simply replaces the LSB of the randomly selected pixel, whereas in LSBM, the value of the pixel is randomly increased or decreased by one to match the message bit. LSBM is also known as ±1 steganography method. HUGO and MiPOD are modern adaptive embedding methods which embed the message based on certain criteria rather than randomly. In HUGO, the message bits are embedded in noisy regions of the image using a Syndrome-Trellis code-based distortion function. MiPOD is a technique that exploits a locally estimated multivariate Gaussian cover image model.

686-dimensional SPAM feature set is used, and the KELM algorithm is implemented using Matlab Software (release 2013b). The experiments are conducted on a machine having specifications as: Dual Intel Xeon Processor E5 2620 v3 CPU with`32 GB RAM running on 64-bit Windows 2008 Server R2 Operating System.

Stego images at 20%, 50%, and 90% embedding are created with all the above embedding techniques. The stego image dataset thus created is divided into two non-overlapping subsets: 50% is used for training and the other 50% is reserved for testing. The feature set computed from the training set is used to train KELM with Gaussian kernel. The KELM hyper parameters, C and γ are optimized using grid search and 5-fold cross-validation. This trained KELM model is then used to classify cover and different stego images from the reserved test dataset.

The experiments conducted include binary and multi-class classification. To ensure the robustness and consistency, the experiments have been conducted 10 times and the results provided are average of these experiments. Figure 10.3 presents the multi-class classification results for stego images created with different payloads and using different embedding algorithms.

We observe from Figure 10.3 that KELM categorizes the classical spatial domain embedding algorithms with high accuracy. The adaptive methods are also classified fairly well. At the most representative payload of 0.5 bpp, we are able to classify the HUGO and MiPOD with accuracies in the range of 70%.

[1] The original idea of Least Significant Bit replacement seems to be due to Romana Machado, who developed software "Stego" for Macintosh systems. An archive of this software is available at https://people.math.osu.edu/fiedorowicz.1/PGP/stego.html

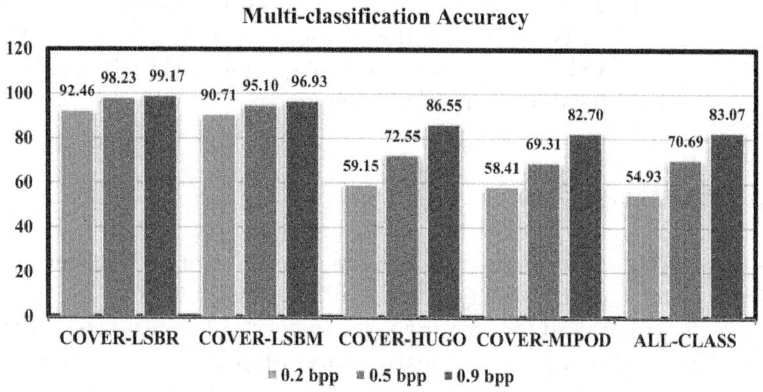

Figure 10.3 Binary and multi-class classification accuracy of KELM.

Table 10.1 KELM classification confusion matrix at different payloads

		Cover (%)	*LSBR (%)*	*LSBM (%)*	*HUGO (%)*	*MiPOD (%)*
0.2 bpp	Cover	36.35	8.68	9.61	23.81	21.54
	LSBR	0.99	85.69	10.68	1.45	1.19
	LSBM	1.60	12.42	81.94	2.18	1.86
	HUGO	20.96	8.81	10.00	39.74	20.48
	MiPOD	23.26	9.05	10.68	26.09	30.92
0.5 bpp	Cover	55.83	2.11	5.75	17.62	18.68
	LSBR	0.34	97.42	1.52	0.45	0.27
	LSBM	1.10	2.63	93.80	1.38	1.09
	HUGO	15.93	2.17	6.03	58.81	17.06
	MiPOD	20.45	2.15	7.62	22.18	47.60
0.9 bpp	Cover	76.22	0.76	3.09	8.58	11.34
	LSBR	0.22	98.74	0.56	0.26	0.22
	LSBM	1.04	1.00	95.76	1.14	1.06
	HUGO	8.67	0.87	3.49	77.69	9.28
	MiPOD	13.40	0.94	5.95	12.79	66.93

Within the binary classification, KELM performance shows significant results even at a reduced payload of 0.2 bpp. The classical embedding methods are rather remarkably classified within the embedding range of 0.2–0.9 bpp. Taking note of the fact that HUGO was specifically designed to elude SPAM features, KELM does a worthy job. The other adaptive algorithm, MiPOD, is also reasonably identified when searching through five different methods.

Furthermore, Table 10.1 presents the confusion matrix for all-class classification results at different payloads. The classical embedding methods are classified with high accuracy even at lower payloads. Moreover, most of the misclassifications between LSBR and LSBMR are inter se.

Intuitively, the adaptive algorithms are misclassified either among themselves or as cover. Nonetheless, these results corroborate the fact that KELM using SPAM features can competently multi-classify classical as well as adaptive steganographic algorithms.

10.6 CONCLUSIONS

This paper presents a multi-class classifier which not only differentiates between cover and stego images but also determines the type of steganographic algorithm used for creating stego images. The multi-class classifier uses SPAM as a feature set. The features are computed by using the second order Markov features of residual images along different directions. The experimental results demonstrate that the proposed steganalysis method effectively classifies images as cover or stegos created with different classical as well as adaptive steganographic techniques. The proposed work can be extended to test KELM with more advanced features sets upon target stego images created with different adaptive algorithms at even lower payloads.

REFERENCES

Böhme, R. 2010 Principles of modern steganography and steganalysis, 11–77. In *Advanced Statistical Steganalysis*. Springer, Berlin, Heidelberg.

BOSS Web page. 2010. http://agents.fel.cvut.cz/boss/index.php?mode=VIEW&tmpl=about, accessed 20 December 2018.

Chandramouli, R. 2003. A mathematical framework for active steganalysis. *Multimedia Systems* 9: 303–311.

Cho, S., J. Wang, C.C. J. Kuo and B.H. Cha. 2010. Block-based image steganalysis for a multi-classifier, 1457–1462. In *The Proceedings of the IEEE International Conference on Multimedia and Expo*, IEEE, Suntec City, Singapore.

Dong, J., Wang, W. and T. Tan. 2009. Multi-class blind steganalysis based on image run-length analysis, 199–210. In A.T.S. Ho, Y.Q. Shi, H.J. Kim and M. Barni [eds.] *Digital Watermarking*, Lecture Notes in Computer Science. Springer, Berlin, Heidelberg.

Harmsen, J.J. and W.A. Pearlman. 2003. Steganalysis of additive-noise modelable information hiding, 131–142. In E.J. Delp and P.W. Wong [eds.] *The Proceedings of Security and Watermarking of Multimedia Contents V*. SPIE, Santa Clara, CA.

Huang, G.B., Zhou, H., Ding, Z. and R. Zhang. 2012. Extreme learning machine for regression and multiclass classification. *IEEE Transactions on Systems, Man, and Cybernetics, Part B (Cybernetics)* 42: 513–529.

Huang, G.B., Zhu Q.Y. and C.K. Siew. 2006. Extreme learning machine: Theory and applications. *Neurocomputing* 70: 489–501.

Lubenko, I. and A.D. Ker. 2011. Steganalysis using logistic regression, 78800K, vol. 7880. In N.D. Memon, J. Dittmann, A.M. Alattar and E. J. Delp III [eds.] *The Proceedings of Society of Photographic Instrumentation Engineers.* Media Watermarking, Security, and Forensics III. International Society for Optics and Photonics.

Petitcolas, F.A.P., Anderson, R.J. and M.G. Kuhn. 1999. Information hiding-A survey, 1062–1078. In *The Proceedings of the IEEE.* IEEE, New York.

Pevný, T. and J. Fridrich. 2005. Towards multi-class blind steganalyzer for JPEG images, 39–53. In M. Barni, I. Cox, T. Kalker and H.J. Kim [eds.] *The Proceedings of the Digital Watermarking. Lecture Notes in Computer Science.* Springer, Berlin, Heidelberg.

Pevný, T. and J. Fridrich. 2007. Merging markov and DCT features for multi-class JPEG steganalysis, 650503. In E.J. Delp and P.W. Wong [eds.] *The Proceedings of the Security, Steganography, and Watermarking of Multimedia Contents IX.* SPIE, Santa Clara, CA.

Pevný, T., Bas, P. and J. Fridrich. 2010a. Steganalysis by subtractive pixel adjacency matrix. *IEEE Transactions on Information Forensics and Security* 5: 215–224.

Pevný, T., Filler, T. and P. Bas. 2010b. Using high-dimensional image models to perform highly undetectable steganography, 161–177. In R. Böhme, P.W.L. Fong and R. Safavi-Naini [eds.] *Information Hiding.* Lecture Notes in Computer Science. Springer, Berlin, Heidelberg.

Rodriguez, B.M. and G.L. Peterson. 2008. Multi-class classification fusion using boosting for identifying steganography methods, 697407. In B.V. Dasarathy [eds.] *The Proceedings of the Multisensor, Multisource Information Fusion: Architectures, Algorithms, and Applications.* SPIE, Orlando, FL.

Sedighi, V., Cogranne, R. and J. Fridrich. 2016. Content-adaptive steganography by minimizing statistical detectability. *IEEE Transactions on Information Forensics and Security* 11: 221–234.

Trivedi, S. and R. Chandramouli. 2003. Active steganalysis of sequential steganography. In III E.J. Delp andP.W. Wong [eds.] *The Proceedings of the Society of Photographic Instrumentation Engineers 5020, Security and Watermarking of Multimedia Contents V. International Society for Optics and Photonics.*

Veena, S.T. Arivazhagan, S. and W.S.L. Jebarani. 2019. Improved detection of steganographic algorithms in spatial LSB stego images using hybrid GRASP-BGWO optimisation, 89–112. In J. Hemanth and V. Balas. [eds.] *Nature Inspired Optimization Techniques for Image Processing Applications.* Intelligent Systems Reference Library. Springer, Cham.

Xu, X., Sun, Y., Wu, J. and Y. Sun. 2018. Steganography algorithms recognition based on match image and deep features verification. *Multimedia Tools and Applications* 77: 27955–27979.

Zhu, J., Guan, Q. and X. Zhao. 2015. Multi-class JPEG image steganalysis by ensemble linear SVM classifier, 470–484. In Y.Q. Shi, H. Kim, F. Pérez-González and C. Yang [eds.] *The Proceedings of the Digital-Forensics and Watermarking.* Lecture Notes in Computer Science. Springer, Cham.

Chapter 11

An efficient key management solution to node capture attack for WSN

P. Ahlawat and M. Dave

National Institute of Technology Kurukshetra

CONTENTS

11.1 INTRODUCTION

Wireless sensor networks (WSN) comprise a set of large resource-inhibited sensor nodes having wide range of applications. These applications may range from industrial monitoring, environmental control, target tracking, smart homes, agriculture, healthcare etc. [1]. Due to its small memory and processing capability, heavyweight security protocols cannot be applied to secure the ongoing transmission between the sensor nodes. The adversary can physically capture the sensor node and extract the keys. It uses these cryptographic keys to mount other attacks in the network such as the sybil attack and clone attack in network. It is observed that the efficiency of any cryptographic security protocol directly depends on an effective key management scheme (KMS).

Physical node compromise is one of the well-known attacks in WSN. This paper addresses the minimization of key exposure problem during a node compromise. The proposed scheme splits the key into different key slices in order to increase the security during a predistribution server. After splitting the key slices, hash-based mechanisms are used to compute the derivative versions of key slices [3,4]. To further enhance the security, the unused key slices are erased from node key ring [5]. Deployment in hostile areas makes the sensor nodes prone to various adversarial attacks. With the stolen symmetric keys, the adversary may compromise the communication of other valid network nodes. Thus, how to design a secure key management solution without creating much overhead on the sensor nodes in terms of storage, communication, and computation is a major challenge. To mitigate the effect of node capture on the other nodes, we propose an effective key management solution. It assigns the key slices instead of complete key which further reduces the key leakage in the network during a node capture. The major highlights of the proposed work are given below:

i. To increase the resilience of KMS, hash chaining is used on the predistributed keys. Due to the one-way characteristic of the hash function, the keys are concealed in such a way that derivative versions of the key do not reveal the original keys.
ii. To further enhance the resistance against node capture of the proposed KMS, a key splitting method is used that predistributes the key slices instead of the original keys. It results in less key leakage to an adversary during a node compromise.
iii. After the link key is established in shared key discovery phase, the sensor nodes delete the key slices that are not used for link key establishment. This enhances the robustness against adversarial attacks of the proposed KMS.

The paper is structured as follows: The overview of different schemes in WSN security is discussed in Section 11.2. System models and definitions are given in Section 11.3. Section 11.4 describes the proposed scheme. Security analysis is done in Section 11.5. We conclude the paper in Section 11.6.

11.2 RELATED WORK

Numerous symmetric KMSs are proposed to secure the ongoing communication among sensor nodes. In pairwise key distribution, every node shares an exclusive pairwise key with its neighboring nodes in the network [2]. This scheme ensures that every pair of node has a unique pairwise key establishment. This scheme has perfect resilience to adversarial attacks. But it has a huge storage overhead on small resource-constrained sensor nodes. It also suffers from scalability issues. The most feasible KMS schemes are

the probabilistic schemes that are random in nature. The earliest probabilistic key predistribution scheme is given by Eschenauer and Gligor called EG scheme [6]. Sensor nodes can set up a secure link key only when they have at least one common key identifier. It was later enhanced by Chan et al. in Reference [7], in which it was proposed that at least q common keys are required to overlap instead of just one key. This scheme has better resilience than EG scheme, but scalability was the issue. Moreover, it does not provide node authentication. This scheme fails in large-scale attacks. Anita et al. [8] merged the polynomial-based scheme and q-composite scheme to generate the pairwise key between sensor nodes. It is shown that this scheme has higher possibility of key connectivity and security with reduced overhead. The concept of isolated nodes and their significance in key connectivity is given in Reference [9]. Some schemes use deployment knowledge to enhance the local connectivity among the sensor nodes [10]. The only limitation of such scheme is requirement of prior deployment information. Hash function used to improve the security of predistribution phase in a KMS is presented by Bechkit et al. [11]. In this scheme, keys of a node are hashed depending on the node identifier value. Thus, if an adversary captures a sensor node, it can discover the keys of the nodes having node identifier greater than the compromised node. The process of computing the actual resilience of a network is given in Reference [12]. Another scheme to minimize the node capture impact by collision keys is given by Jiri Kůr et al. [13]. Wenqi Yu [4] reduced the effect of compromised nodes on the communication between uncompromised nodes using hash function. Shan and Liu [14] proposed the use of hash function to reduce the information that is revealed out when a node gets compromised. Another technique to increase the communication security is the concept of key splitting. In this method, a single key is divided into different subkey parts and these parts are stored in sensor nodes. It enhances the security of the resulting scheme without degrading other parameters. A key deletion process presented by Ehdaie et al. [5] states that every node should discard the unused key from their key ring after link key establishment is over in a network. However, to tackle this problem a new solution is also given by authors. It does not have any storage overhead on the sensor nodes. This scheme also has no communication or computation overhead. Hence, the resilience of this scheme is increased as compared with other schemes. Ma et al. [15] proposed a KMS for heterogeneous WSN. An analysis on the resilience of q-composite scheme is presented by Zhao [16]. Various scheme parameters are derived and optimal values of these parameters are obtained that defend against adversarial attacks. Adversarial-based KMS is presented in Reference [17] to increase the security strength of overall network. A symmetric design-based hybrid key distribution is presented by Anzani et al. [18] to improve the connectivity and security. A parameter is defined based on application requirement to provide a good tradeoff between security and connectivity of the proposed KMS. A review on different attacks, challenges, and

security solutions is given by Sharma and Vaid [19]. It aims to provide analysis of different possible threats in WSN and their proposed defensive mechanisms.

Thus, there is always a need for such a scheme that can increase the security strength of the network without degrading other parameters. The problem of key exposure during a node compromise its minimization is one of the focus points of this paper.

11.3 SYSTEM MODEL AND PROBLEM DEFINITION

In this section, we present different models and definitions used in the proposed scheme. Table 11.1 presents various symbols and notations used in the proposed scheme.

11.3.1 Link key

A link key is fully compromised if all its constituent subkeys are completely exposed to the adversary during a physical node capture. It is said to be partially compromised if complete set of keys that are used to make the link key is captured by the adversary. If some of its keys are captured, then link key is said to be partially compromised.

11.3.2 Threat model

An adversary is said to have limited computational power. It captures the nodes randomly to get the cryptographic keys stored in them. If a node is physically compromised by the adversary, then its keys get open to the adversary. The adversary then uses this keying information to compromise other non-compromised links in the network. To fully devastate WSN security, the secret symmetric keys have to be extracted [3,4]. If adversary wants to compromise all links, then it should get all the cryptographic keys used to constitute these link keys.

Table 11.1 Symbols and their meaning

Symbol	*Meaning*
N	Total number of network nodes
P	Size of key pool
L	Key size in bits
M	Size of the key ring
m'	Key ring size after key erasure
A	Fraction of keys retained by node after key erasure
ADV	Adversary benefit factor
Z	Key splitting factor

11.3.3 Network model

The network is presented as a graph in which the vertex represents a sensor node and edge depicts the link between sensor nodes. The network uses random key distribution.

11.3.4 Hash function

A hash function is a function which when applied to an input, outputs another value. It has a property that we cannot get the original versions of the value from its derived values. The second property states that from any input, the hash function never outputs the same hashed value. This property is used to increase the security of key distribution scheme in which the predistributed keys are hashed before storing into the sensor nodes. We cannot calculate the original keys from their derivative versions. Thus, if a derivative key is captured by the adversary, then adversary cannot generate the original key and hence, the resilience of the KMS is increased against the node capture attack [4].

11.3.5 Key splitting method

This method is used to increase the resilience of the KMS. The key is divided into key slices before predistributing it in the sensor nodes keeping other parameters same. The key server splits the $|P|$ keys present in the key pool into $z \times |P|$ key slices. The benefit of this method is less key leakage. In this method, adversary is likely to hold the complete portion, hence less key information is exposed to an adversary, which thereby diminishes the node capture impact on the KMS [3] (Figure 11.1).

11.4 PROPOSED SCHEME

To obscure the predistributed keys of the sensor nodes a hash function is applied before the key predistribution phase. It aims to provide resistance to the network by applying hash-based mechanisms to the predistributed key slices. The proposed scheme has three phases, namely key pool generation, which is an offline phase, key assignment, and link key establishment. The basic EG scheme has key ring size defined by the key pool size and

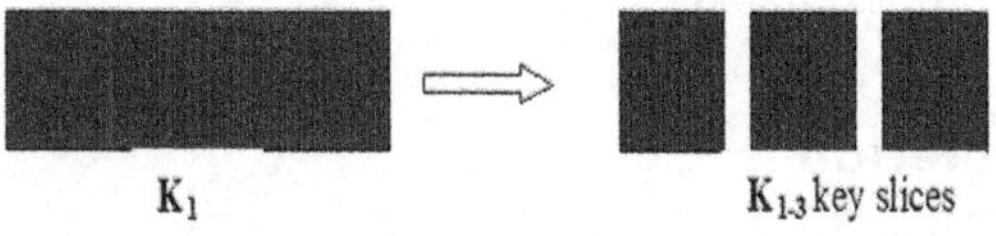

Figure 11.1 Process of key splitting.

network size resulting in poor scalability and storage overhead. The phases of the proposed scheme are explained below.

11.4.1 Key pool generation

In this phase, the key server computes a large key pool *P*. Each key is identified by its unique identifier number. The key server then applies hash-based mechanism on the keys. The original keys are mixed with their derivative versions thereby doubling the size of original key pool, that is, $2P$. Now, the original key pool consists of $2P$ keys with original keys along with their derivative versions. After that, each key is split into z equal key slices. Thus, resulting in a key pool of $z \times 2|P|$. If the size of each key is length L bits, then size of the resulting key slices becomes $L/|z|$ bits. The memory requirement of storing m keys will now become $m.|L|$ bits.

11.4.2 Random key assignment

In this phase, the key server randomly assigns $m \times z$ key slices to the sensor nodes. After the sensor nodes are placed into the network, hash function is also stored in the sensor nodes along with key identifiers. It is assumed that the original keys and derivative keys have same key identifiers. Bit value is used to differentiate the original keys from their derivative versions.

11.4.3 Shared key discovery

To discover the shared key, the nodes exchange key identifiers with their neighboring nodes. In basic random key distribution, the nodes attempt to locate common key identifier to set up the secure link. In the proposed key management solution, the nodes aim to discover the common key slices to establish the common link. During a node capture attack, an adversary aims to compromise the nodes, and compromise of a single cryptographic key can reveal the link established between the non-compromised nodes. With the same captured nodes, the probability of compromising the secure link of non-compromised nodes in the proposed scheme is decreased by storing the key slices instead of the complete key. It is less likely to hold all the secret key slices to complete the original key. A bit value is also sent to indicate whether it has original version of the key or its derived version. The sequence of key slices is used to know the number of key slices that are common. In the proposed scheme, small key slices are generated from a larger-sized cryptographic key.

11.4.4 Key deletion

Once the shared key discovery is over, all nodes have found their corresponding key pairs. The node erases the key slices that are not used for link

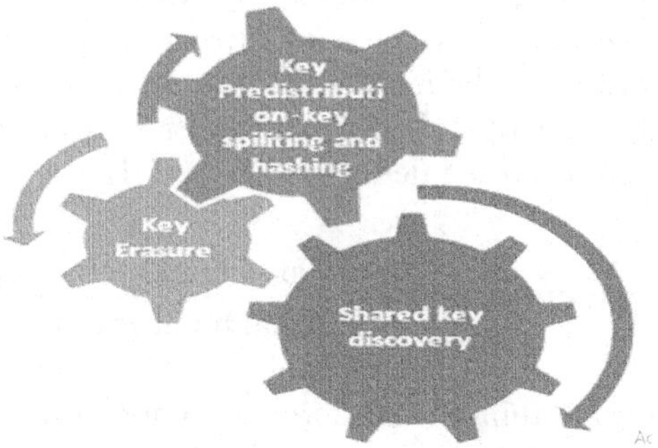

Figure 11.2 Phases of the proposed scheme.

setup. This is done in order to further reduce the node capture impact. It is due to the fact that the capturing of unused key slices may reveal the secure link of other non-compromised nodes.

The nodes broadcast their key identifiers to locate common keys with neighbors they share a key. They also broadcast the type of key along with m number of key identifiers. Suppose nodes have common key identifier, namely n_a and n_b, two cases arise:

a. Nodes n_a and n_b have either original key slices or both nodes have derivative versions of required key slices, say g, then they establish the link key if $g>z$. The sensors use a hash function h to get a secret key as $h(k_1\|k_2\|\ldots\|k_j)$ or $h(k_1'\|k_2'\|\ldots\|k_j')$, where || is a concatenation function, and $k_1, k_2, \ldots, k_j$ are the number of shared key slices of a key k.
b. Node n_a has original key slices and node n_b contains the derivative version of key slice. In that case, node n_a performs hashing on its original key slices and vice versa (Figure 11.2).

11.5 SECURITY ANALYSIS OF THE PROPOSED SCHEME

This section presents the security analysis of the proposed scheme based on the adversary benefit parameter. During a node capture attack, an adversary captures a large number of nodes to get the secret keys. Links of non-compromised nodes become vulnerable if their link key is incorporated in the captured key set. Due to node compromise, a portion of link key of non-compromised node may get revealed to the adversary. We deal with fully compromised links. Suppose the adversary has got sufficient number of key slices to generate the complete key. One of prime concerns is how to increase the security of random key predistribution, thereby reducing the

links compromised. We further assume that there is no reparative action in the network. We assume a constant rate of attack. As the number of captured nodes increases, the number of links of non-compromised nodes get compromised. We have taken a parameter to analyze the additional number of links compromised defined by Eq. (11.1) as:

$$\text{Adversary benefit parameter (ABP)} = \frac{\text{Number of links compromised}}{\text{Total number of network links}}. \quad (11.1)$$

In EG scheme, the probability to compromise a non-compromised link during a node capture is $m/|K|$, where m is the number of total keys assigned to node, and P is total number of keys.

Similar to analysis in references [3,5], the portion of compromised links of a WSN, given a node is captured by the adversary, is given as:

$$\text{ABP}(1) = \Pr[A] = \Pr[A|B].\Pr[B] + \Pr[A|B']\Pr[B]. \quad (11.2)$$

The possibility of capturing any two link edges is given by $\Pr[B]=2/n$, resulting in the probability of not compromising any two link edges equaling to $1–2/n.\Pr[A|B]$, and it becomes 1. The proposed scheme applies hash function to generate the derivate version of original key slices. This increases the resistance of proposed scheme, and capturing of derivative key slices cannot reveal the original ones. However, the compromise of original keys may reveal the derivative versions of the key slices. The final key pool consists of different key slices in which half of the key slices are original and half are derivative. Hence, the probability that m key slices are compromised during a node capture becomes $3.m/4.|K|$. Putting these values in Eq. (11.2), we get the following values:

When n reaches to infinity, we have

$$\frac{1.2}{n} + \frac{3.m}{4.|K|} = \frac{2}{n} + \frac{3.m}{4.|K|}. \quad (11.3)$$

As key splitting is used, each duo of sensor node uses two half key slices. Thus, fail function using this method becomes $\Pr[B]=2/n$. It results in probability of not capturing any two link edges to $1–2/n.\Pr[A|B]$, and it also becomes 1.

The probability of non-compromised nodes getting revealed is only if two half key slices are obtained by the adversary, that is, $(2.3.m/2.4.P)^2 = (3.m/4.p)^2$. Finally, we get the modified values of Eq. (11.3) as

$$\frac{2}{n} + \left(\frac{3.m}{4.p}\right)^2. \quad (11.4)$$

To further enhance its security, we erase the unused key slices after the shared key discovery. As the key slices are randomly distributed in the sensor nodes, these are erased from node key ring. Hence, only m' keys remain in a sensor node. If the adversary captures a node, only $m' \times z$ key slices are revealed to an adversary compared with original $m \times z$ key slices. Thus, the adversary benefit parameter to compromise latest secured links between valid nodes using Eq. (11.4) becomes [3]:

$$\frac{2}{n}+\left(\frac{\alpha.3.m}{4.p}\right)^2, \tag{11.5}$$

where α is the fraction of keys retained by the sensor nodes. It is equal to $(m' \times z)/(m \times z)$.

The comparative analysis of different KMSs in terms of fail function is given in Table 11.2.

To analyze different KMS, we have presented Table 11.2. The EG scheme has lowest value of fail function as it does not have any hash-based mechanism. Wenqi scheme is improved over the EG scheme by introducing hash mechanism. The possibility of key leakage gets reduced by derivative versions of original keys. Bechkit scheme improves the EG scheme using node identifiers–based hash. It suffers from smart attack in which attacker captures the low identifier nodes. Key splitting scheme method improves the fail function but it does not use hash function. Key deletion also reduced the fail function by removing the unused keys. We observe that the proposed scheme has least value of fail function due to three factors, namely key splitting, hash chaining, and key deletion.

Table 11.2 Comparative analysis of different KMSs

Scheme	*Fail function*
EG scheme [6]	$\frac{m}{p}$
Wenqi Yu [4]	$\frac{3.m}{4.p}$
Bechkit scheme [11]	$\frac{2l.m}{l+1.p}$, where l is $h=i$ MOD L
Key splitting scheme [3]	$\frac{2}{n}+\left(\frac{m}{p}\right)^2$
Proposed scheme	$\frac{2}{n}+\left(\frac{3.m}{4.p}\right)^2$
Key deletion [5]	$\alpha.\left(\frac{m}{p}\right)$, where α is $\frac{m'}{m}$

To analyze the adversary benefit parameter, we have taken three schemes, namely the EG scheme, key splitting scheme [3], and proposed scheme as given in Figure 11.3.

To plot Figure 11.3, we have taken the key ring size as 100, key pool as 1,000, and the number of network nodes is assumed to be 100. From the figure, it is shown that proposed scheme has least value of adversary benefit parameter when compared with other schemes. It is because the proposed scheme applies one-way hash function on the key slices which results in increased security of the proposed scheme. Further the security is improved using the key erasure of unused key slices after shared key discovery is over.

To further analyze the effect of variation in the key ring size on adversary benefit parameter, we plot Figure 11.4.

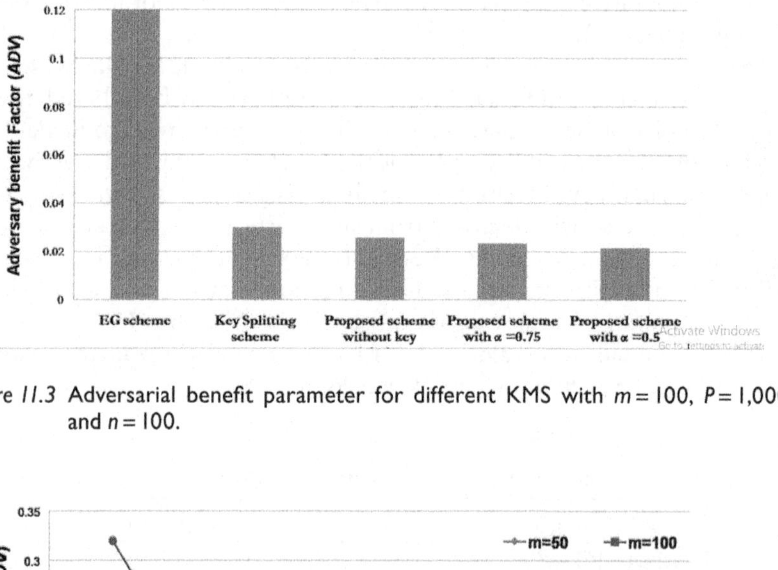

Figure 11.3 Adversarial benefit parameter for different KMS with $m = 100$, $P = 1{,}000$, and $n = 100$.

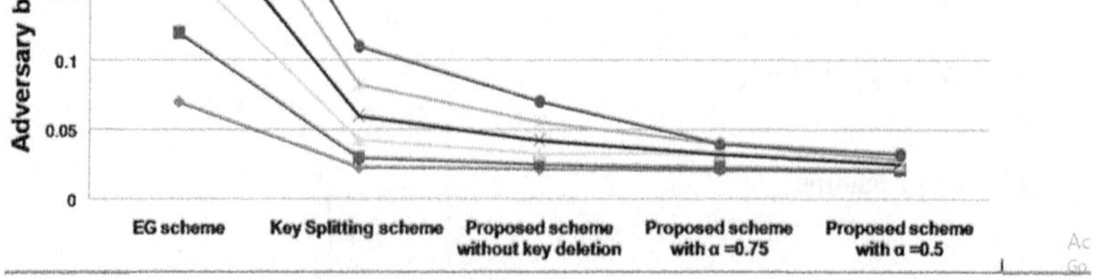

Figure 11.4 Adversarial benefit parameter for different KMS with different values of m keeping $P = 1{,}000$ and $n = 100$.

In Figure 11.4, we observe that even if the keys of the sensor nodes are increased, the value of adversary benefit parameters remains least as compared with EG and key splitting scheme. The derivative key slices do not reveal the original key slices. Hence, in the proposed scheme there is less key leakage. It proves that proposed scheme is more secure than other schemes. To study the effect of variation in key pool size on the value of adversary benefit parameter, we plot Figure 11.5.

In Figure 11.5, we find that with the increase in key pool size the value of adversary benefit parameter decreases. This is because the probability of getting the key slices by the adversary during a node capture gets decreased with increase in pool size. So, when we increase the key pool size, the value of adversary benefit parameter is least in proposed scheme. It proves that proposed scheme has better resilience as compared with other schemes.

In Figure 11.6, we observe that by increasing the number of nodes, the adversary benefit parameter increases. It is due to the fact that large secret

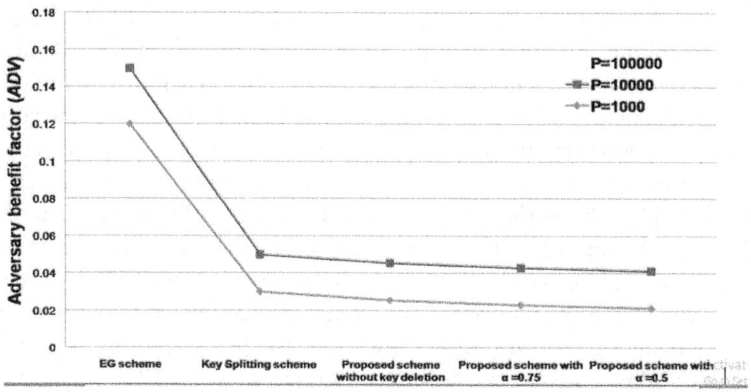

Figure 11.5 Adversarial benefit parameter for different KMS with different values of *P* keeping *m* = 100 and *n* = 100.

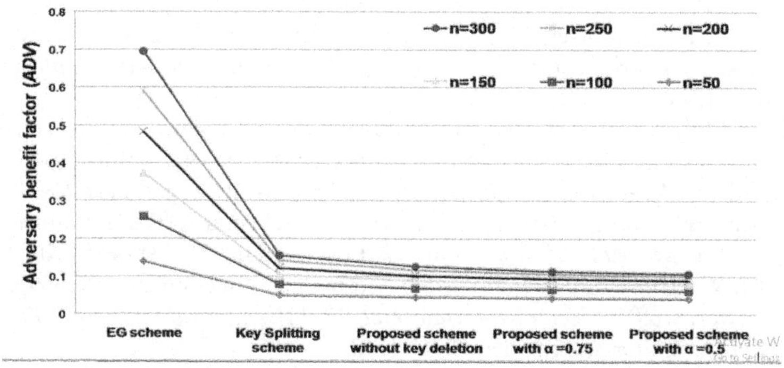

Figure 11.6 Adversarial benefit parameter for different KMS with different values of *n* keeping *m* = 100 and *P* = 1,000.

keying information gets exposed to an adversary in a node compromise. It is smallest in proposed scheme as compared with other existing schemes. As the proposed scheme uses hashed key slices to set up the link key, the security of overall network is increased. Deletion of unused key slices from the node memory after shared key discovery further reduces the value of adversary benefit parameter. It proves that proposed scheme has better resilience as compared with other schemes.

11.6 CONCLUSION

The security of established link between the sensor nodes becomes a critical issue as the sensor nodes are placed in unattended environments. This may result in large capturing of sensor nodes, thereby reducing the resilience against node capture of KMS. To mitigate the effect of node capture on the communication links of the sensor nodes, in this paper, we presented a secure key management solution that is resistant to node capture. Key splitting technique is combined with hash chaining to reduce the possibility of link key compromise by an adversary. With same number of captured nodes, the proposed scheme reveals lesser keying information from compromised node than basic random key distribution. In addition, the security of the proposed scheme is further strengthened by the process of key deletion that deletes the unused key slices from the sensor nodes. This lowers the possibility of link compromise during a node capture. The obtained results have proved that the proposed scheme is able to increase the security of the KMS by lowering the probability of link comprise between non-compromised nodes in terms of adversary benefit factor. It is also shown that the proposed scheme performs better than other schemes for different values of n, m, and P.

REFERENCES

1. He, X., Neidermeier, M., Meer, H.: Dynamic key management in wireless sensor network: a survey. *Journal of Network and Computer Applications*, 36, 612–622 (2013).
2. Aikyildiz, I.F., Su, W., Sankarasubramaniam and Cayir, E.: Wireless sensor networks: a survey. *Computer Networks*, 38(4), 393–422 (2002).
3. Ehdaie, M., Alexiou, N., Attari, M. A., Aref, M. R., Papadimitratos, P.: Key splitting: making random key distribution schemes resistant against node capture. *Security and Communication Networks*, 8(3), 431–445 (2015).
4. Yu, W. (2010). A pairwise KMS based on hash function for wireless sensor networks. *2010 Second International Workshop on Education Technology and Computer Science* (198–201). IEEE.
5. Ehdaie, M., Alexiou, N., Ahmadian, M., Aref, M. R., Papadimitratos, P.: Mitigating node capture attack in random key distribution schemes through key deletion. *Journal of Communication Engineering*, 6(2), 99–109, 2017.

6. Eschenauer, L., Gligor, V.: A key-management scheme for distributed sensor networks. *Proceedings of 9th ACM Conference on Computer and Communications Security*, Association for Computing Machinery, New York, NY, United States, 41–47 (2002).
7. Chan, H., Perrig, A., Song, D.: Random key predistribution schemes for sensor networks. In *2003 Symposium on Security and Privacy*, 2003. pp. 197–213, IEEE.
8. Anita Mary, E.A., Geetha, R., Kannan, E.: A novel hybrid KMS for establishing secure communication in wireless sensor networks. *Wireless Personal Communications*, 82, 1419–1433 (2015).
9. Gupta, B., Pandey, J.: Non existence of isolated nodes in secure wireless sensor networks. *Wireless Personal Communications*. DOI: 10.1007/s11277-015-2845-9 (2015).
10. Du, W., Deng, J., Han, Y., Chen, S., Varshney, P.K.: A KMS for wireless sensor networks using deployment knowledge. *Proceedings of IEEE INFOCOM'04*, IEEE, 586–597 (2004).
11. Bechkit, W., Challal, Y., Bouadallah, A.: A new class of hash chain based key predistribution scheme for WSN. *Computer Communications*, 36, 243–255 (2013).
12. Yum, Hyun D., Lee, P. J.: Exact formuale for reselience in random key predistribution schemes. *IEEE Transactions on Wireless Communications* 11, 1638–1643 (2012).
13. Kůr, J., Matyáš, V., Švenda, P.: Two improvements of random key predistribution for wireless sensor networks. In *International Conference on Security and Privacy in Communication Systems*, 61–75, Springer, Berlin, Heidelberg (2012).
14. Shan, T. H., Liu, C. M.: Enhancing the key pre-distribution scheme on wireless sensor networks. *Asia-Pacific Services Computing Conference, 2008. APSCC'08. IEEE*, 1127–1131, IEEE (2008).
15. Ma, C., Shang, Z., Wang, H., Geng, G.: An improved KMS for heterogeneity wireless sensor networks. *International Conference on Mobile Ad-Hoc and Sensor Networks*, 854–865, Springer, Berlin, Heidelberg (2007).
16. Zhao, J.: On resilience and connectivity of secure wireless sensor networks under node capture attacks. *IEEE Transactions Information Forensics and Security*, 12(3), 557–571 (2017).
17. Ahlawat, P., Dave, M.: An attack model based highly secure KMS for wireless sensor networks. *Procedia Computer Science*, 125, 201–207 (2018).
18. Anzani, M., Javadi, H. H. S., Modirir, V.: Key-management scheme for wireless sensor networks based on merging blocks of symmetric design. *Wireless Networks*, 24(8), 2867–2879 (2018).
19. Sharma, C., Vaid, R.: Analysis of existing protocols in WSN based on key parameters. In *Proceedings of 2nd International Conference on Communication, Computing and Networking*, 165–171. Springer, Singapore (2019).

Chapter 12

Privacy preservation and authentication protocol for BBU-pool in C-RAN architecture

Byomakesh Mahapatra, Awaneesh Kumar Yadav, Shailesh Kumar, and Ashok Kumar Turuk
National Institute of Technology, Rourkela

CONTENTS

12.1 INTRODUCTION

The growth of internet of things (IoT) and other data-centric wireless network leads to the increase in traffic load on the base stations (BS). To encompass this increasing load the number of the cellular BS should be increased, which again leads to an increase in cost incurred and network complexity (CISCO 2016). Cloud radio access network (C-RAN) is considered to be an alternative to overcome these limitations. The C-RAN architecture reduces the associated cost to a larger extent by using the cloud and virtualization technology at the RAN platform. C-RAN provides a collaborative RAN platform for accessing different baseband signals over a virtualized baseband unit (V-BBU) (Wu et al. 2015). The C-RAN architecture (Figure 12.1)

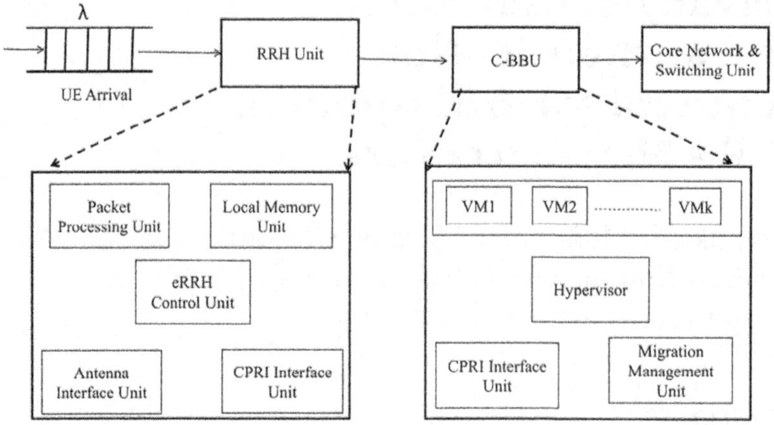

Figure 12.1 Block diagram representation of C-RAN architecture.

consists of a distributed radio unit also known as remote radio head (RRH) where the number of user entities (UEs) is linked by sharing a wireless channel. The baseband signal received by the RRH unit is again down converted and sent to a centralized unit known as a baseband unit (BBU), where all the signal processing and controlling activity is carried out. The centralized BBU (C-BBU) is grown over a cloud platform which consists of physical hardware, virtual machine (VM), and a hypervisor for performing controlling and interfacing activity. C-RAN allows many cellular operators to use a single platform based on RAN-as-a-service concept by creating their own VMs. This interleaving of many operators leads to data hijacking and data tempering by breaking the integrity and confidentiality of the system. To overcome this issue, the C-BBU needs an advanced authentication and privacy preservation method (Park and Park 2007).

In this chapter, we highlight some of the common security threats to the C-BBU and propose an authentication mechanism to secure the C-RAN platform. As C-RAN needs to authenticate users as well as service providers, it needs a complex authentication and verification procedure. In the proposed method, the UEs are authenticated by the BBUs, whereas the VMs are verified by the corresponding hypervisor of the C-BBUs.

The chapter is organized as follows: Section 12.2 highlights some of the relevant work on C-RAN and conventional data center security. The architecture of C-RAN is described in Section 12.3. The C-RAN advantages and their limitations are presented in Section 12.4. Various security challenges and their impact on C-RAN are described in Section 12.5. The background of the proposed work is presented in Section 12.6. The protocol authentication and verification procedures are described in Section 12.7. The simulation results and corresponding discussion are given in Section 12.8. Finally, concluding remarks are mentioned in Section 12.9.

12.2 RELATED WORK

C-RAN is a new BS architecture which can handle different types of radio frequency (RF) signals on a single platform by using virtualization and cloud computing technology over a BBU. The collaborative radio processing needs an improved security and authentication strategy unlike the traditional data center (Mahapatra et al. 2018). In Tian et al. (2017), authors described different key security challenges in C-RAN and proposed many solutions to overcome these limitations. Ahmad et al. (2018) discussed about the possible attacks in 4G cellular network. The paper proposed different security solutions for some common attacks like distributed denial-of-service (DDoS), denial-of-service (DoS), and man-in-the-middle (MITM) attacks for 4G cellular network. Authors in Mpitziopoulos et al. (2003) discussed about possible attacks and their corresponding security solutions for wireless sensor networks. On the other hand, Xiao and Xiao (2013) and Hyde (2009) gave a complete overview of the security challenges in the cloud computing environment. They have also focused on two main possible attacks, host operating system (OS) attack and guest OS attack, in a virtualized data center.

But all the previous papers were limited to the analysis rather than practical protocol implementation. To fill this void, we have proposed a new authentication protocol for BBU of a C-RAN architecture. Different validation tools validated the protocol after performing the simulation to a number of iterations.

12.3 ARCHITECTURE OF C-RAN

Based on the functionality, the whole architecture of C-RAN is divided into following subunits:

a. *Remote radio head (RRH)*: RRH is an RF unit which sends and receives the uplink/downlink signal through a set of RF channels toward the UE and C-BBU.
b. *Centralized baseband unit (C-BBU)*: C-BBU performs all the control and processing activities of the C-RAN. This unit consists of a number of VMs, which are created with the help of a guest OS. The guest OS or VMs are controlled by another application run over a host OS known as hypervisor or virtual machine manager as shown in Figure 12.2.
c. *Fronthaul and backhaul connection*: The fronthaul and backhaul are wired or wireless type connections, where the former one is used for signal transmission between RRH and BBUs and latter one for signal transmission between BBU and core network.

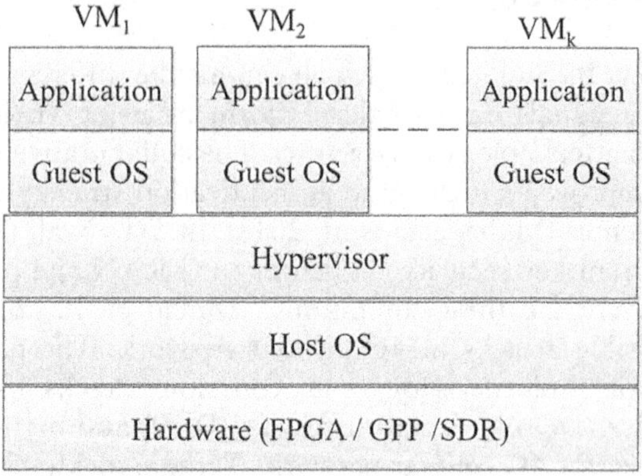

Figure 12.2 Layer architecture virtualized C-BBU.

12.4 ADVANTAGES OF VIRTUALIZATION AT C-BBU

The use of virtualization technique at the C-BBU offers many substantial advantages to the RAN world:

- *Collaborative radio access*: As virtualized BSs use a single platform to provide services to different cellular operators based on infrastructure as a service, this helps to bring many services to a common platform.
- *Services isolation*: The use of different VMs for different cellular operators isolates the services running over a single platform. This isolation also provides a secured baseband processing without interfering with other VM services.
- *Easy maintenance and fast recovery*: As all the baseband signals are processed on one platform at a centralized location, it is easy to maintain all the resources within a short recovery period (Bian and Park 2006).
- *Cost-effective and secured services*: The centralization brings down different costs to a greater extent. Again, the introduction of centralized authentication and key agreement protocol simplifies the VM authentication process (Lin et al. 2010).

12.5 SECURITY CHALLENGES IN VIRTUALIZED C-BBU

Implementation of virtualization in radio access technology adds many security challenges at the C-BBU. The attacker attacks the hypervisor of C-BBU through either host-OS or guest-OS. There are various types of attacks possible at the C-BBU of C-RAN, like DDoS attack, MITM attack,

honeypot attack, and guest-to-guest attack, at both the host and the guest OS. The prime objective of this chapter is to find out different possible attacks and corresponding solutions to overcome these attacks. The different forms of possible attacks at the C-BBU are as follows:

- *Direct host attack*: In this attack, the attacker attacks on the hypervisor by taking advantage of the vulnerability and security holes present in the host OS. In this attack, the main aim of the attackers is to take control over the hypervisor as well as the host OS (Zhange and Jain 2011).
- *DoS attack*: In DoS attack, the attacker can compromise the hypervisor or host OS by sending a large number of VM creation requests using fake user identity. A compromised hypervisor is unable to handle these large number of requests, as it becomes busy in only managing the huge requests instead of serving requests. This creates an unbalanced resource utilization condition at the BBU-pool (Douligeris and Mitrokotsa 2003).
- *Migration attack*: This is another form of MIMT. This attack occurs at the C-BBU during the migration of the VMs from one host to other. The attacker exploits the security protocol and vulnerability of the network and imposes some malicious code to the target VM. When the affected VM enters the BBU-pool, it compromises the whole system and gets access to all the BBU resources (Schneider and Horn 2015).
- *Indirect host attack*: In this attack, the hypervisor gets attacked by an unauthorized user, by creating unauthorized VM over the host OS. Due to the effect of compromised VM, the other VMs present in the same host and controlled by the same hypervisor can also get affected (Sonar and Upadhyay 2014).

12.6 PROPOSED WORK

As the BBUs grow over a data-centric environment with the help of a cloud computing technology, VM and hypervisor securities become prime factors for establishing an end-to-end connection. The proposed privacy preservation and authentication protocol preserves privacy of the VMs and hypervisor for proper resource utilization with the aim of increasing the quality of services of the cellular network. Figure 12.3 shows a complete flow chart for hypervisor key management and BBU utilization in a C-RAN. The whole process is carried out through three different phases mentioned below.

12.6.1 VM request phase

In VM request phase, the user requests the hypervisor to create a VM. The hypervisor analyzes all the circumstances of the host in terms of physical resources available and utilization of each BBU. If the hypervisor finds a

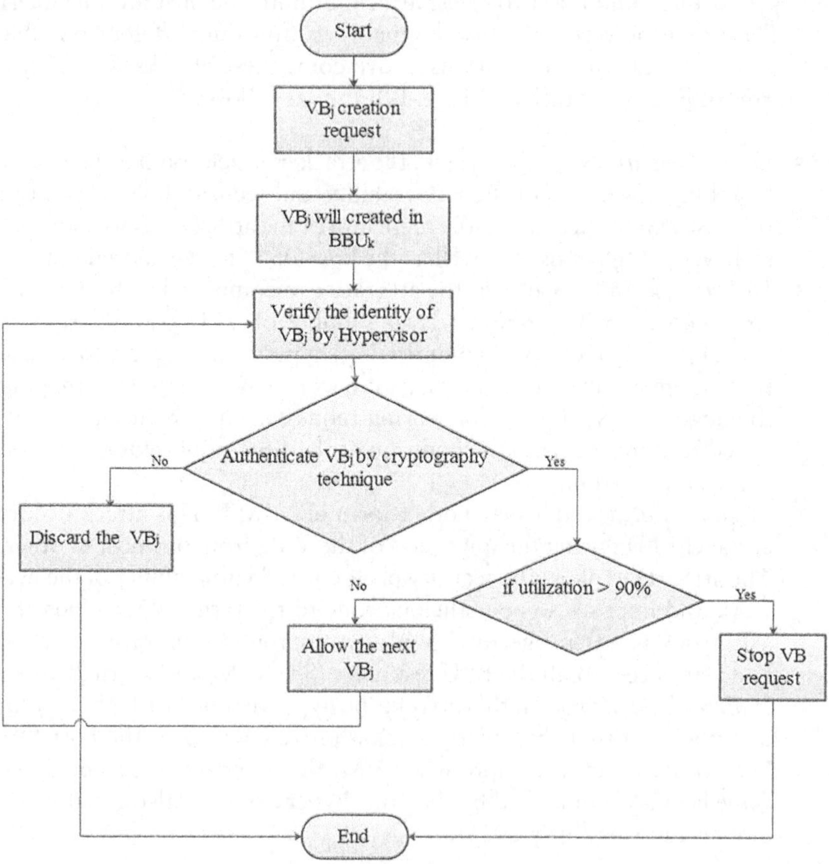

Figure 12.3 Flow diagram of VM assignment and security assessment.

suitable condition for creating a new VM, it accepts the request and starts the authentication procedure for this new request.

12.6.2 VM registration and authentication phase

In this phase, the new VM registers itself to the hypervisor based on the VM allocation policy of the BBU-pool. The registered VM is then authenticated by using our proposed privacy preservation and authentication protocol. This proposed protocol has a key distribution center (KDC) for distributing the keys among the VMs which are associated with the hypervisor. This KDC authenticates the VMs with respect to hypervisor based on the computing challenge response. When the challenge key and the response challenge key of the KDC and VMs are matched, then the VM authentication process is completed.

12.6.3 Host utilization calculation phase

After allocating a specific number of VMs to the host, the hypervisor each time updates the utilization of the BBUs before assigning a new VM to the host. When the host utilization reaches a threshold value, the hypervisor will generate an alert message to the host OS to either stop accepting new VM requests or perform load balancing by using a load balancing technique. During this period, the hypervisor continuously monitors each VM and host for proper resource utilization and authentication.

12.7 PROPOSED SECURITY PROTOCOL VERIFICATION AND AUTHENTICATION PROCEDURE

The proposed protocol is used for both UE and VM authentication and verification processes. The KDC uses certain secret keys for authentication process. These keys are shared by the KDC to UE and VM. A detailed description of the keys used is given in Table 12.1.

12.7.1 Step 1: UE authentication and verification step

In this step, a user requests to create a VM. The hypervisor authenticates the legitimacy of the user request with the help of a KDC function as shown in Figure 12.4. The detailed authentication procedure is as follows:

- If UE or operator wants to create a VM, it will send a VM creation request to the host OS with a unique identity UE_{id}.
- After accepting the request from the UE, the host OS will verify the legitimacy of the user request and send it to the KDC for the verification by sending the identities (UE_{id}, HOS_{id}).

Table 12.1 Symbols and abbreviations

Symbols	*Definition*
KDC	Key distribution center
KDF	Cryptographic function (key derivation function)
UE	User entity
HOS	Host operating system
VM	Virtual machine
Sk	Shared secret key
R	Random number
AUTN	Authentication vector
K_{asme}	MME intermediate key (access security management entity)
RES/XRES	Response/expected response
CH/RCH	Challenge/response challenge

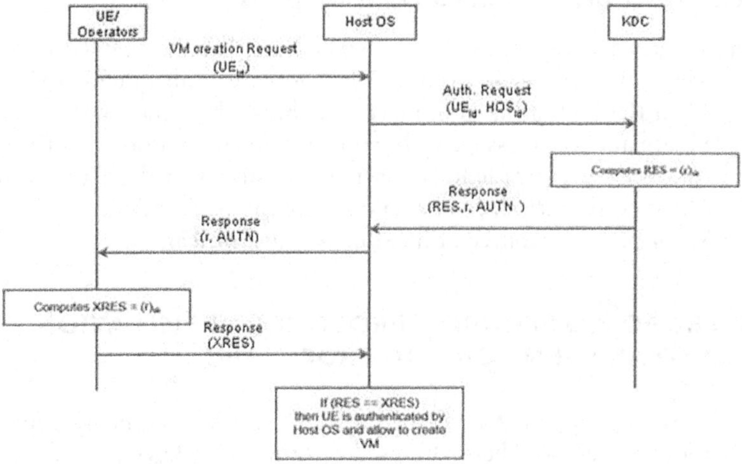

Figure 12.4 Protocol for UE authentication by the host OS with the help of KDC function.

- The KDC will accept the request and compute the response for the verification process with the help of a secret key, shared between UE and KDC:

$$\text{RES} = (r)_{sk}. \tag{12.1}$$

- After computing the RES, KDC will send it to the host OS with (RES, *r*, AUTN), where AUTN is the authentication token that acts as real-time nonce for the verification of session.
- Host OS keeps the computed response for further verification process and sends (*r*, AUTN) to the user.
- After accepting the (*r*, AUTN), UE will also compute the expected response with the help of shared secret key of the user and send it to the host OS for verification.
- Now, host OS will accept the expressed response XRES and verify it with RE:

$$\text{if}\left(\text{XRES} == \text{RES}\right). \tag{12.2}$$

If the secret key shared between users and KDC is same, it means user is authenticated and will be allowed to create a VM, otherwise it will simply reject the VM creation request.

12.7.2 Step 2: VM authentication and verification step

After creating the VMs, the authentication and key preservation process is carried out by the KDC and its associated hypervisor. The authentication process is shown in Figure 12.5. The steps are as follows:

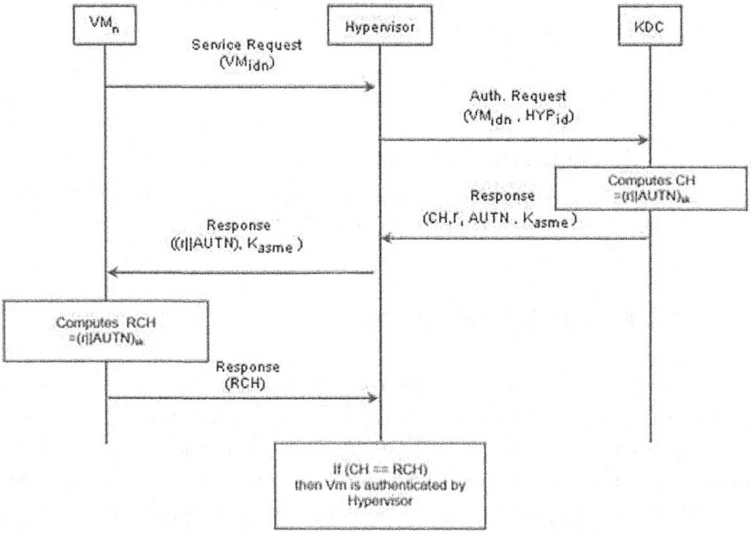

Figure 12.5 Protocol for VM authentication by the host OS with the help of KDC function.

- VM will request for service to the hypervisor by sending identity VM_{idn}.
- After accepting the request from the VM, hypervisor will verify the legitimacy of the VM through the KDC. So, it will send the verification request to the KDC by sending the identities (VM_{idn}, HYP_{id}).
- Now, KDC will accept the request and compute the response for verification with the help of the secret key shared between VM and hypervisor and generate the K_{asme} as calculated for the further process of verification:

$$\text{CH} = (r \parallel \text{AUTN})_{(\text{sk})} \tag{12.3}$$

$$K_{\text{asme}} = \text{KDF}\ (\text{sk}, \text{HYP}_{\text{id}}, r) \tag{12.4}$$

where K_{asme} is access security management entity used to secure the verification process when CH or RCH gets compromised by an attacker.

- After computing the CH, KDC will send it to hypervisor with (r, AUTN, K_{asme}, CH).
- Hypervisor keeps the computed challenge CH and sends (r, AUTN, K_{asme}) to VM.
- After accepting the (r, AUTN, K_{asme}), the VM will also compute the response challenges and send it to the hypervisor for verification.

- After accepting the RCH, hypervisor will verify the challenge value along with the response:

$$\text{if}(\text{CH} = \text{RCH}). \tag{12.5}$$

If secret keys shared between VM and hypervisor are same, it means VM is authenticated and will grant the service request to the hypervisor, otherwise simply suspends the server request.

12.8 RESULT AND SIMULATION

We have simulated the proposed protocol in two security tools: automated validation of internet security protocols and applications (AVISPA) and SCYTHER for the validation and verification (Armando et al. 2005). The protocol simulation is done using AVISPA tool with a system configuration of 4 GB RAM along with 64-bit OS. The proposed protocol is validated using two different simulators.

In the initial phase, the proposed protocol was tested and validated in two modes of the AVISPA tool which are constraint-logic-based attack searcher (CL-AtSe) and on the fly mode (OFMC) as shown in Figures 12.6 and 12.7, for UE and VM authentication procedure, respectively.

The protocol simulation is also carried out with the help of SCYTHER tool in a Linux environment. Figure 12.8 shows that the proposed protocol is unaffected by the possible attacks by securing the VMs and its associated hypervisor.

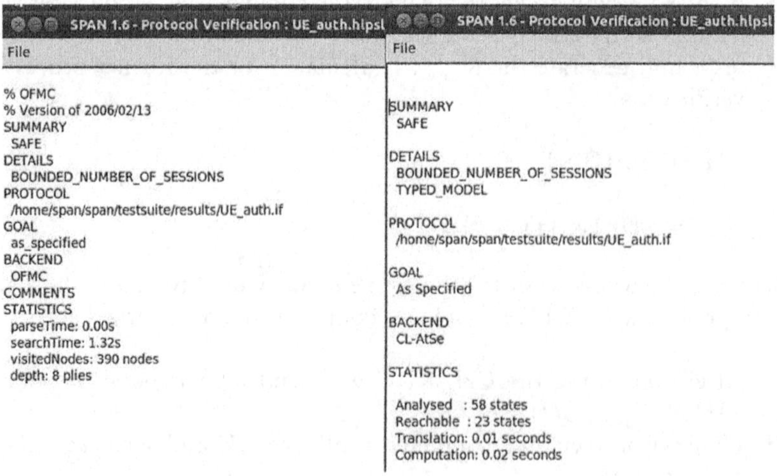

Figure 12.6 UE authentication and verification using CL-AtSe backend and OFMC backend of AVISPA tool.

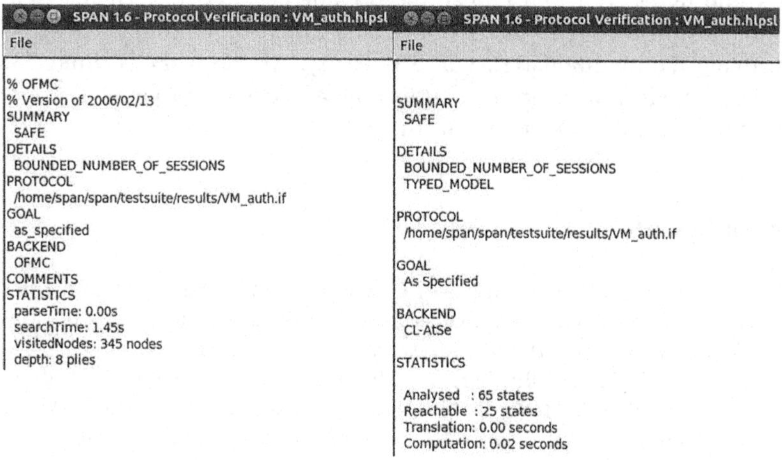

Figure 12.7 VM authentication and verification using CL-AtSe backend and OFMC backend of AVISPA tool.

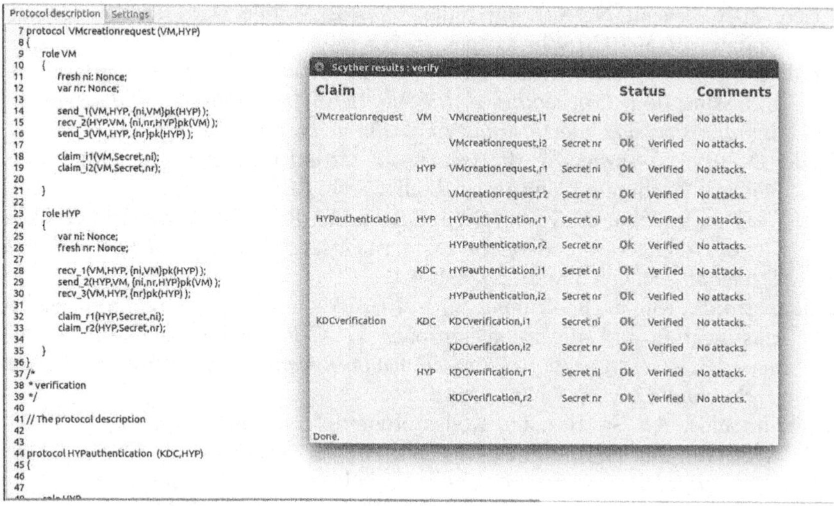

Figure 12.8 VM authentication in SCYTHER tool.

12.9 CONCLUSION

Virtualization and cloud computing technology aim to improve the quality of services by providing a scalable and flexible RAN platform. As C-RAN uses a single cloud platform to provide services to many cellular and network operators, providing VM and hypervisor security is the key challenge for infrastructure provider. The proposed protocol is designed to secure

VMs and hypervisor and reduce the possible attacks to a greater extent. Again, this protocol can be used to secure a real-time cellular network by improving data confidentiality and integrity. As the work is limited to a simulated environment, in near future the authors plan to extend the protocol to real-time RAN environment.

REFERENCES

Ahmad, I., Kumar, T., Liyanage, M., Okwuibe, J., Ylianttila, M., and Gurtov, A. 2018. Overview of 5G security challenges and solutions, *IEEE Communications Standards Magazine*, vol. 2(1), pp. 36–43.

Armando, A., Basin, D., Boichut, Y., Chevalier, Y., Compagna, L., Cullar, J., Drielsma, P. H., Ham, P. C., Kouchnarenko, O., Mantovani, M., and Mdersheim, S. 2005. The AVISPA tool for the automated validation of internet security protocols and applications, *International Conference on Computer Aided Verification*, pp. 281-–285.

Bian, K., and Park, J.M. 2006. MAC-layer misbehaviors in multi-hop cognitive radio networks, *US-Korea Conference on Science, Technology, and Entrepreneurship (UKC2006)*, pp. 228–248.

Cisco, 2016. Visual Networking Index, Global Mobile Data Traffic Forecast Update, 2015–2020 White Paper.

Douligeris, C., and Mitrokotsa, A. 2003. DDoS attacks and defense mechanisms: a classification, *Proceedings of the 3rd IEEE International Symposium on Signal Processing and Information Technology*, pp. 190-–193.

Hyde, D. 2009. A survey on the security of virtual machines. Dept. of Comp. Science, Washington Univ. in St. Louis, Tech. Rep.

Lin, Y., Shao, L., Zhu, Z., Wang, Q., and Sabhikhi, R. K. 2010. Wireless network cloud: Architecture and system requirements, *IBM Journal of Research and Development*, vol. 54, (1), pp. 4:1–4:12.

Mahapatra, B., Kumar, R., Kumar, S., and Turuk, A. K. 2018. A real time packet classification and allocation approach for C-RAN implementation in 5G network, *4th International Conference on Recent Advances in Information Technology (RAIT)*, Dhanbad, pp. 1–6.

Mpitziopoulos, A., Gavalas, D., Konstantopoulos, C., and Pantziou, G. 2003. A survey on jamming attacks and countermeasures in WSNs, *IEEE Communications Surveys & Tutorials*, vol. 11(4), pp. 42-–56.

Park, Y., and Park, T. 2007. A Survey of Security Threats on 4G Networks, *IEEE Globecom Workshops*, pp. 1–6.

Schneider, P., and Horn, G., 2015. 5G Security, *IEEE Trustcom/BigDataSE/ISPA*, Helsinki, pp. 1165–1170.

Sonar, K., and Upadhyay, H. 2014. A survey: DDoS attack on Internet of Things, *International Journal of Engineering Research and Development*, vol. 10(11), pp. 58-–63.

Tian, F., Zhang, P., and Yan, Z. 2017. A survey on C-RAN security, *IEEE Access*, vol. 5, pp. 13372–13386.

Wu, J., Zhang, Z., Hong, Y., and Wen, Y. 2015. Cloud radio access network (C-RAN): a primer. *IEEE Network*, vol. 29(1), pp. 35–41.
Xiao, Z., and Xiao, Y. 2013. Security and privacy in cloud computing, *IEEE Communications Surveys Tutorials*, vol. 15(2), pp. 843–859.
Zhang, M., and Jain, R. 2011. Virtualization security in data centers and clouds, In http://www.Cse.Wustl.Edu/~jain/index.Html.

Chapter 13

Threshold-based technique to detect a black hole in WSNs

Chander Diwaker and Atul Sharma
Kurukshetra University

CONTENTS

13.1 INTRODUCTION

Wireless sensor networks (WSNs) are often required to detect, handle, and spread information of centered physical circumstances. WSNs contain battery-operated sensor gadgets with enrolling, data getting ready, and passing on fragments. The manners in which the sensors are passed on can either be in a controlled space where watching and perception are fundamental or in an uncontrolled circumstance. In the uncontrolled circumstances, security for sensor frameworks ends up being critical [1,2].

The advancement of WSNs as one of the dominating development designs in the recent decades has posed different unprecedented troubles for scientists. Its identifying development joined with planning vitality and wireless affiliation makes it rewarding for being man-handled in riches in future. The introduction of wireless affiliation development also causes distinctive sorts of security causes. The objective of this paper is to look at the security-related issues of WSNs and to propose solutions to secure the WSNs against the security threats [3,4]. WSNs are creating prerequisite of clients. It has such a large number of highlights over the customary network and shortcomings as well. We need to protect a huge number of devices while planning specially appointed networks. The premier undertaking

of the specialists is to make impromptu networks safe. There are endless attacks in WSNs. A short description of all attacks is given: black hole attack is one of the significant attacks [5,6]. It is an attack in which nodes quietly expel or drop the entering or leaving messages without illuminating the source that message did not get at its goal. These nodes remain imperceptible in the network and are recognized just when activity lost is checked. This paper focuses of these kinds of attacks.

Scientists have done parcel of work on this attack. The researchers proposed numerous calculations to identify and evacuate the black hole attack. It is attractive that all the proposed plans give different strategies and procedures to manage black holes, yet nobody gives a protected and vitality productive component [7–9].

This paper is categorized into five sections: Section 13.1 presents the introduction of WSNs. Section 13.2 presents the literature review of various existing techniques related to black hole detection in WSNs, and Section 13.3 provides proposed threshold-based black hole detection system model. Section 13.4 gives the results and analysis of implemented proposed work, also the results are compared with existing routing scheme.

13.2 RELATED WORK

Jatav et al. [8] displayed an instrument to dispatch sinkhole-based attacks, for example, particular sending and black hole attack in WSNs. The paper incorporates the usage of sinkhole attack on mint-route convention in WSNs. Recreation results demonstrate that these attacks were increasingly serious. The number of parcels at base station diminished exponentially. This work incorporates findings and countermeasures to make the sensor network secure from these kinds of attacks. It was commonly distinguished that different proposed techniques for recognition and countermeasure accomplished a high level of security with unimportant overheads.

Sheela et al. [9] proposed a lightweight, speedy, viable, and flexible pro-advancement which relied upon the security plan against dim attack in WSNs. WSN has a dynamic topology, which comprises the spasmodic network and resources obliged of the diverse contraption center points. This proposed arrangement was used to secure against dull opening attack using distinctive base stations passed on in the framework by diverse convenient authorities. A package drop attack or dull hole strike could be seen as a kind of denial of organization ambush wrapped up by dropping bundles. These attacks can be mastered by explicitly dropping packages for an explicit framework goal or by a package every *t* seconds, by a self-assertively picked piece of the packages, which was designated "diminish hole ambush", or in mass by plummeting all groups. The adaptable administrator is seen as an undertaking section with all things considered as self-controlled. The investigation begins with one central point and then goes onto the following

center points in the midst of transmitting data. In addition, doing computation simultaneously was seen as an amazing perspective for the coursed applications, and generally engaging in a dynamic framework condition. This segment generally does not require greater imperativeness. Here, it is realized that using a reenactment-based model of the response, the system can recover from the diverse black hole attacks in a WSN. The relationship between correspondence overhead and the cost was established among the systems without utilizing different base stations. The system with diverse base stations tends to turn away dim opening attacks. The examination was also done between the proposed strike disclosure structures using a flexible administrator against the security system without the convenient experts. The adaptable authorities were delivered using the Aglet.

Ramachandran et al. [10] checked the effects of dull hole and flooding strike on mobile ad hoc networks (MANETs) and WSNs by reenacting the dim opening and hustling attack in the geographic multicast routing. The simulation was generally done using NS-2. The framework execution was done keeping all factors unchanged with and without the dull hole and hustling strike in the WSN, and a relationship was established with MANET. The package transport extent, throughput, end-to-end deferral, and imperativeness hardship have been evaluated and on appraisal a reduction was noticed in these components. In dim hole strike, generally, all the framework traffics were occupied to the specific center point or from the noxious center making this real mischief the geographic multicast routing tradition. In hustling ambush, because of the more drawn out transmission lines in each center point to get the execution, the framework is ruined. From the execution estimations, it was fathomed that the dim opening was an outrageous coordinating ambush for WSN.

Dighe et al. [11] proposed different strategies that were utilized to successfully relieve the unfavorable impacts of the black hole attacks on the WSNs. The system was organized with different base stations in the network while directing multiple information bundles to these base stations. The arrangement was exceedingly compelling and needed next to no computation and message trade in the network, thus sparing the vitality of the sensor networks. The reproduction results given by them demonstrate that this plan accomplishes 99% bundle conveyance achievement and 100% black hole node recognition.

Yadav et al. [12] introduced a fluffy based choice to check whether a node was tainted by a black hole attack or not. The authors designed an interruption location framework to distinguish the black hole attack on Ad-hoc On-demand Distance Vector (AODV) in MANETs. This recognition framework was based on fuzzy logic. The fluffy standard was actualized on the reaction time of node communication. Presently, as opposed to exchanging information on this node, it will pass on the information from the surrounding nodes; it will just contract with the communication that was synchronized to it as it were.

13.3 PROPOSED MECHANISM TO DETECT BLACK HOLE NODES IN WSNs

The present research work focuses on the rank-based scheme to detect black hole nodes in WSNs. In this section, the proposed work is presented in detail.

13.3.1 Proposed algorithm

```
Start
  N=Nodes in network
  for (i=0; i <= n; i++)
  {
     Compute Rank information of each Node
        ((msgid_source=msg_received by Node) && (msg_received by
Node=msg_forwarded by Node)

        If Rank Ni >= 0.5
  {        Then node status is set="Honest Node"
  Else
  Node status="Black Hole"
     }
  }
  Stop
```

13.3.2 Description of proposed mechanism

The proposed framework focuses on ruined, that is, counterfeit data or genuine data of every node. This data are determined based on checking the message created by the sending node or receiving the same message by transitional node. Additionally, if same message is sent by halfway node to another node, then node status is not a black hole. These data are used to process rank of the node. The value of rank of each node is in the range of 0 to 1. So, the threshold value is set to 0.5, as it is the optimal value to distinguish black hole nodes from normal nodes.

In the event that the node is having rank less than settled esteem (threshold esteem), it is recognized as ordinary node, else the node is a black hole. This calculation is difficult to utilize and execute. The proposed plan can be executed in a completely appropriated way to defeat different attacks without depending on any carefully designed equipment. The weakness of the previously mentioned conventions is mimicked utilizing opportunistic network environment (ONE) simulator. ONE is an open source simulator that can be run on Linux as well as Windows operating system. It is a java-based simulator.

13.3.3 Data Flow Diagram (DFD) for the proposed solution

Figure 13.1 describes the flow of working of the proposed approach using a flow chart.

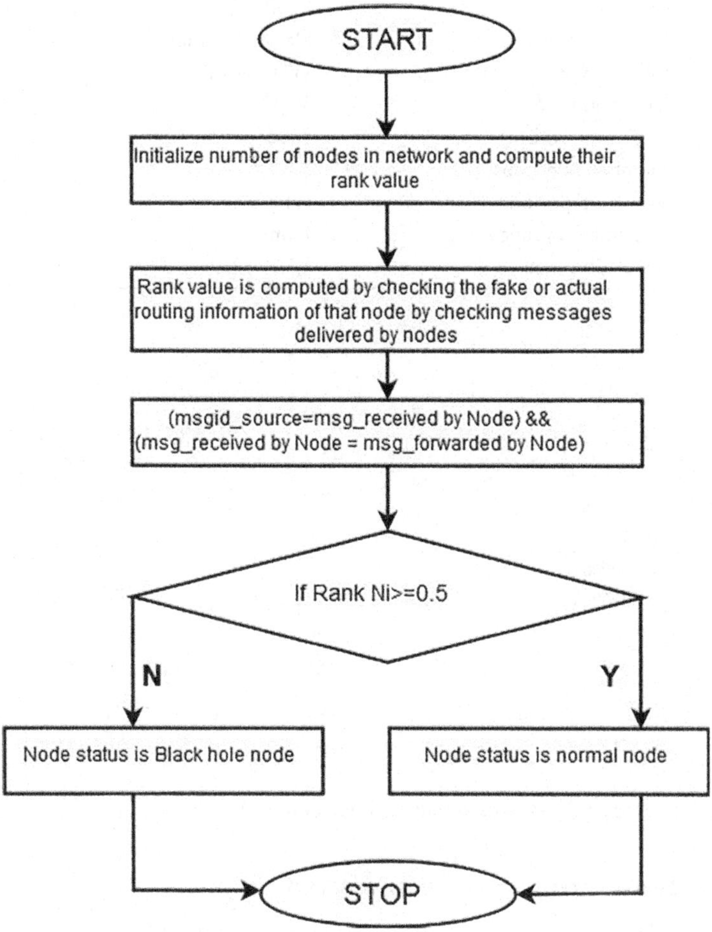

Figure 13.1 Flow chart of the proposed mechanism.

13.4 RESULTS AND ANALYSIS

Table 13.1 shows simulation parameters used to simulate the proposed model. The parameters such as mobility model, buffer size, message size, time to live, number of groups, transmission range, number of nodes have been used for simulation of the proposed model.

Table 13.1 Parameters used in simulation

Parameter description	*Value*
Routing	Prophet routing
Number of nodes	[50;100;150;200]
Message size	500 KB–1 MB
Buffer size	5 MB
Mobility model	Map-based movement model
Operating system used	Windows 7
Simulation time	30,000 s
TTL (time to live)	300 s
Number of groups	3
Warm-up period	1,000 s
Transmission range	10 m
Simulation area	$4{,}500 \times 3{,}400\,m^2$

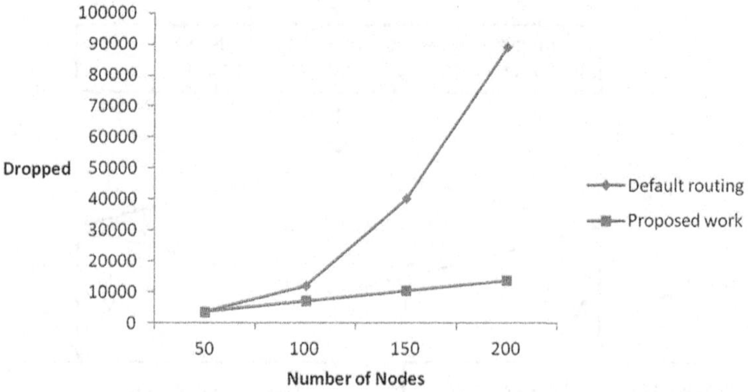

Figure 13.2 Dropped messages versus number of nodes.

13.4.1 Simulation results and analysis

After simulation, reports module generates the results. The following graphs are obtained by considering aforementioned network metrics in contrast with the increasing number of nodes. The results of routing with the proposed algorithm and routing without black hole nodes are compared.

Figure 13.2 exhibits the effect of a number of dropped messages on existing and proposed scheme at variable nodes. It is clearly visible from the graph that when black holes are present in the network, then packet drop rate is very high and message drop rate decreases abruptly in proposed scheme, that is, the drop rate in proposed scheme is low. In the proposed

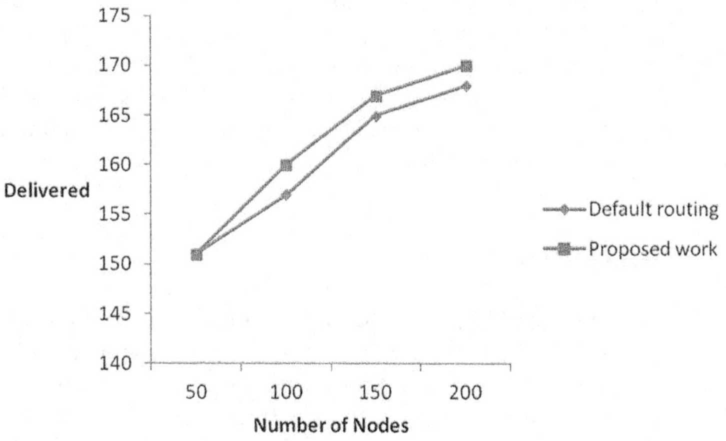

Figure 13.3 Messages delivered versus number of nodes.

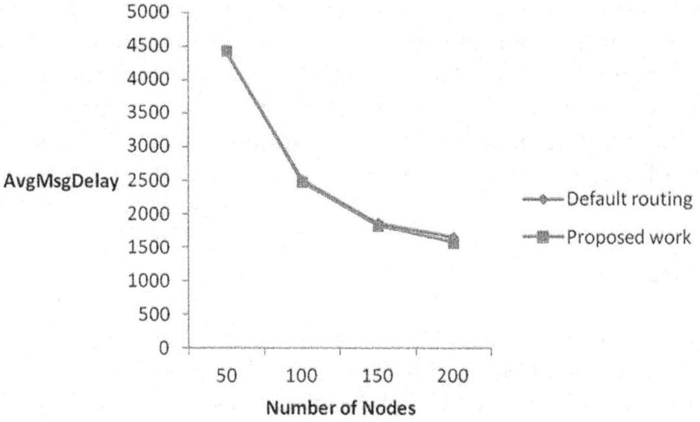

Figure 13.4 Average message delay versus number of nodes.

work, packet drop rate is decreased to 85% than in the existing routing. So, we can say that the proposed scheme is efficient in reducing the drop rate.

Figure 13.3 shows the impact of a number of messages delivered on the existing and proposed scheme. It is clearly evident from the graph that message delivery rate increases as we increase the number of nodes. The proposed scheme delivers greater number of packets as compared with the existing scheme. Packet delivery rate in the proposed scheme is slightly greater than the existing routing. Delivery rate varies from 150 to 170. As nodes increase, delivery rate also increases.

Graphical results in Figure 13.4 show the average message delay rate with increasing number of nodes. A message delay is a time taken for the messages from their creation to delivery. It is clearly noticeable that there is a slight drop in average delay in our proposed scheme as compared with the existing scheme. In the existing scheme, routing delay is high, but in the proposed scheme it is improved.

13.5 CONCLUSION

In this paper, a threshold-based procedure is proposed. A threshold is a settled incentive by which we contrast conveyance proportion of every node for recognizing black holes. Moreover, we propose a plan to check whether a node is ruined or genuine by steering calculation which is also examined. We have examined a portion of the essential parts of WSNs and showed how network execution can be enhanced with map-based movement model. Factors like conveyed, dropped, throughput, average message postpone measurements are utilized for results examination. To recreate the proposed work, ONE test system is utilized. Reenactment results demonstrate that the proposed plan is enhanced than past plans in the context of use, asset utilization, and so forth. In this examination it is observed that the proposed scheme has gained huge ground in anchoring WSNs. In future, it is expected to keep chipping away at it and propose another dependable effective plan with the utilization of encryption code to keep the network away from black holes.

REFERENCES

1. Shreenath K. N., Manasa V. M. (2017) Black hole attack detection in zone based WSN. *Int. J. Recent Innovation Trends Comp. Comm.* 5(4): 148–151.
2. Kaurav A., Kakelli A. K. (2017) Detection and prevention of black hole attack in wireless sensor network using NS-2.35 simulator. *Int. J. Sci. Res. Comp. Sci. Eng. Info. Tech.* 2(3): 717–722.
3. Guptha Y. P. K., Madhu M. (2017) Improving security and detecting black hole attack in wireless sensing networks. *Int. J. Prof. Eng. Stud.* 8(5): 260–265.
4. Otoum S., Kantarci B., Mouftah H. T. (2017) Hierarchical trust based, black hole detection in WSN based smart grid monitoring. *IEEE ICC Ad-Hoc and Sensor Networking Symposium*: 1–6.
5. Chhabra A., Vashishth V., Sharma D. K. (2017) A game theory based secure model against black hole attacks in opportunistic networks. *IEEE Info. Sci. and Sys. (CISS)*, Baltimore: 1–6.
6. Hemalatha P., Vijithaananthi J. (2017) An effective performance for denial of service attack (dos) detection. *IEEE Int. Conf. on I-SMAC (IoT in Social, Mobile, Analytics and Cloud)*, Palladam: 229–233.

7. Aljumah A., Ahanger T. A. (2017) Futuristic method to detect and prevent black-hole attack in wireless sensor networks. *Int. J. Comp. Sci. Net. Sec.* 17(2): 194–201.
8. Jatav V. K., Meenakshi Tripathi, Gaur M. S., Vijay Laxmi (2012) Wireless sensor networks: Attack models and detection. *IACSIT Hong Kong Conf.* 30(2012): 144–149
9. Sheela D., Srividhya V.R., Asma Begam, Anjali and Chidanand G.M. (2012) Detecting black hole attacks in WSNs using mobile agent. *Int. Conf. Arti. Intel. Emb Sys.* 1–5.
10. Ramachandran S., Shanmugam V. (2012) Performance comparison of routing attacks in MANETs and WSN. *Int. J. Ad hoc, Sensor Ubiq. Comp.* 3(4): 41–52.
11. Dighe P. G., Milind, Vaidya B. (2012) Deployment of multiple base stations to counter effects of black hole on data transmission in WSN. *Int. J. Eng. Inno. Tech.* 1(4): 209–213.
12. Yadav P., Gill R. K., Kumar N. (2012) A fuzzy based approach to detect black hole attack. *Int. J. Soft Comp. Eng.* 2(3): 2231–2307.

Chapter 14

Credit card fraud detection by implementing machine learning techniques

Debachudamani Prusti, S. S. Harshini Padmanabhuni, and Santanu Kumar Rath
National Institute of Technology Rourkela

CONTENTS

14.1 INTRODUCTION

Credit card fraud is a very pertinent problem and the way of fraudulent source of funds in an online transaction throughout the credit card industry [1]. Of late, various researchers, as well as applicationists, have shown interest in the analysis of fraud issues in the credit card by applying machine learning algorithms. In real situations, it is necessary to respond in a very short time to stop fraudulent transactions. Additionally, due to the varying

behavior of fraudulent methods, a frequent re-training is essential for any credit card fraud detection model.

Still the major troublesome problem to handle the counterfeit is that detecting the fraud for credit cards is a cost-sensitive issue, in which the expense delivered by a false alarm is not quite the same as the cost of a false negative class [2]. When the model predicts an online transaction as fraudulent, but actually it is not (false positive), the organization has both an administrative cost and a large decline in consumer satisfaction. In addition, when the model is able to identify a fraudulent transaction (false negative), unconditionally the loss occurs for that transaction. Most importantly, it is not sufficient to have a fixed cost variance in between false positives as well as false negatives, as the total cost of the transactions vary in a significant way. Hence, its financial impact is not fixed.

In this study, a classification algorithm model has been proposed to classify the financial transactions to be either fraudulent or not. A good number of computational and statistical methods such as Bayes classifier, nearest neighbor, discriminant analysis, and logistic regression have been proposed in the literature to develop various classification models for the accurate prediction of accuracy value [3]. Other artificial neural networks and classification-based trees, artificial intelligence, and machine learning techniques were also applied for the classification task [4,5]. In view of risk control, estimation of default will be more informative as compared with classifying the customers into fraudulent and non-fraudulent. Hence, the major issue lies in estimating the probability of defaults produced from different data mining techniques that represent the actual probability of defaults.

A class of stolen or lost cards has turned into a noteworthy issue, that is, the stealing of cards from the e-mail. This is also known as non-receipt issue (NRI) fraud that impacts issuers during issue of both new card as well as reissue of duplicate cards. Some specific geographic areas of the nation are at more risk than other areas for NRI. In other places, this issue is so severe to the point that the issuers utilize some substitute strategies for card delivery, that is, courier, with some unique activation procedure for the credit cards. With this, a card remains blocked till the customer calls the bank to check and approve the card receipt. When the bank receives the call, it validates that the customer is the approved cardholder by making inquiries assumed from the acknowledged cardholder's application or information file. Hence, such projects have prompted decrease in NRI misfortunes.

Some other varieties of frauds include submission of fake applications to a bank in order to get credit cards. In this situation, culprits attempt to acquire one's data and other important certificates. They use this information to send a mail to the address the card is ought to be sent. If mistakenly a card is issued in such scenarios, then even the concept of card member activation will not stop the card going into the hands of frauds. Other

sources in fraudulent activity are through mail or telephone order fraud [4]. In such situations, at the time of the transaction, the buyer is not available in front of the merchant where no card imprint is obtained as a transaction record. Efforts to combat this type of fraud have included address verification check via telephone with the account holder during the transaction.

There are also some frauds that originate from the merchant's end. Such frauds include the illegal delivery of receipts, acquiring large amount of money for the transaction that has not happened earlier. Likewise, the merchant's hidden agreement results in the merchant's establishment, that is, utilized as an area where the account data are imitated amid the time of legal transactions. These data are used to deliver a new duplicate card, which is later utilized somewhere else for fraudulent transaction activity. In this situation, the merchant is viewed as a point of middle ground; all such fake duplicate cards have histories involving transactions that can be followed back to use at the given merchant's location.

Again, there is an area of fraud which is mainly referred as abuse [6]. For this situation, the cardholder buys using the card without having any intention to repay. In other cases, this premeditated crime action happens before the cardholder's petitioning for individual bankruptcy. Misfortunes because of bankruptcy fraud issue are not viewed as a part of the credit card counterfeit issue itself, with charge off losses.

14.2 CREDIT CARD FRAUD ISSUE

Fraudulent activities in credit card include smart techniques of stealing the identity cunningly where the unauthorized person employs some other person's credit card credentials for the payment of purchasing anything or transfer of amounts from the cardholder's account [6]. The fraud in credit card can also have an impact with the fraudulent transaction of debit card which can be exploited by stealing the original card. It uses the cardholder's account information and the credit card credentials like card number, pin number, name, and address unauthorizedly.

Irrespective of the fact that credit card takes numerous structures, there are various methods of important classification techniques [7]. Lost cards and stolen cards are the measure dimensions of counterfeit where most of the parts represent a specific misrepresentation action of credit card. The measure of this base dimension can be influenced by general monetary conditions. Fraud because of fake cards has turned into a current developing issue in the course of recent years, in spite of taking preventive measures for more modern card producing innovations (3D hologram or images on the card) and the sensitive data encryption on the stripe. Clearly, fraud will in general be an increasingly sorted out and deliberate issue in specific areas, instead of the more artful and subsequently determined nature of most counterfeit because of lost or stolen cards.

14.2.1 Current methods of fraud detection

The decent variety of fraud action as affirmed by many forms of counterfeit is a complex task for the detection of deceitful behavior of any transaction [5]. Almost all the financial institutions have some variety of scrutiny procedure to apply for new credit cards including routine check of data to keenly identify the probable deceitful applications. Sometimes, the scrutiny process of the application forms for obvious strategies for handwriting has driven investigators to spot fake applications put together by sorted out criminal components. When the credit card has been issued to the applicant, most banks depend upon occasional verification of account activity to decide whether there is confusion of fraud. Specifically, banks have built up a list of guideline-based verifications in which all records action are looked into. Such investigations may determine the limit the number of transactions which ought to sensibly be required to take place in a particular day. Such an excessive transaction report may likewise be constrained to a number of transactions value beyond some threshold value on the purchased amount.

These counterfeit rules are taken in light of recorded investigations of past deceitful behavior in the portfolio. Many banks utilize just only the essential of measurable analyses to restructure the counterfeit rules, driving as a rule to decide sets that comprise of many straightforward boundary conditions on record factors. Specifically, when visualized as an issue in pattern classification, the problem of fraud identification is an appropriate application for the neural network system solution. Progressively, various issues in financial transactions are being visualized in terms of pattern classification problems for which the neural network system solutions and machine learning solutions might be developed.

14.3 APPLICATION OF MACHINE LEARNING MODELS FOR FRAUD DETECTION

In machine learning algorithms, classification technique is regarded as an occurrence of supervised learning, that is, training where a learning set of accurately viewed perceptions is fully accessible. The corresponding unsupervised learning is called clustering and includes gathering information into classifications dependent on some proportion of intrinsic similarity or separation. Our study considers the following classifiers.

14.3.1 Naïve Bayes classifier

Naïve Bayes classifier works on Bayes theory of conditional probability [8]. It assumes that the attributes are independent to each other (naïve), which is not true in reality because there always exists dependency between the attributes. It gives hypothetical support to different classifiers that

implicitly utilize Bayes hypothesis. The naïve assumption decreases the computational complexity because of the class conditional independence.

14.3.2 Extreme learning machine

Extreme learning machine (ELM) algorithms are similar to the feedforward neural network systems, basically applied for classification, clustering, regression, compression, sparse approximation, and feature learning [9]. It can be implemented using only one layer or multiple layers of hidden nodes. The parameters of each hidden node are not required to be associated with each other. These neural network models provide a generalized performance value and the learning process is faster than systems that are trained by utilizing backpropagation.

Generally, all parameters in the feedforward neural network are often tuned in a manner where the dependency exists between various layers of parameters (i.e., loads and biases). For the past few years, gradient descent-based techniques are being utilized in different learning algorithms like feedforward neural systems. In any case, obviously gradient descent-based learning techniques are commonly eased back because of ill-advised steps of learning which can effectively combine to reach the local minimum.

14.3.3 K-nearest neighbor

The classifier K-nearest neighbor (KNN) is used for classification problems which are dependent on learning algorithms by analogy [10]. For a given unknown sample, the algorithm looks for the KNNs in the sample space. Distance is calculated in the form of separation. The test sample is allotted to the nearest recognized class from its KNNs. The advantage of this methodology is that we do not need to build a model before the classification. The disadvantages include non-delivery of straightforward classification probability formula, and the proportion of separation and the cardinality k of the area exceptionally influence the predictive accuracy.

14.3.4 Multilayer perceptron

Multilayer perceptron (MLP) is also similar to a feedforward neural network used for classification and regression problems [11]. An MLP consists of minimum of three layers: an input layer, one or more hidden layers, and an output layer. Apart from the input layer nodes, other two nodes are the neurons that utilize a nonlinear activation function (NAF). MLP uses a supervised learning system known as backpropagation. It is used for training purpose. Its various layers and NAFs differentiate MLP from the linear perceptron. The nonlinearly separable data can be differentiated by it. In MLP, all the neurons have a linear activation function, that is, a linear function which points out the weighted contributions to the output layer. In MLP, a few neurons utilize NAF developed to display the recurrence of

activity possibilities, or terminating, of organic neurons. For MLP to work well, a large dataset is often required.

14.3.5 Support vector machine

Support vector machine (SVM) is a statistical learning strategy having a wide application in classification algorithms [12]. SVM classification algorithm helps to develop a hyperplane and maximize the separation between the positive and negative modes of classification data volume. Kernel representation and margin optimization are the two crucial properties that help SVM boundary separation effective.

This model finds a very uncommon sort of linear model, the sharp-edge hyperplane, and it classifies all training data instances very efficiently by classifying them into right classes through a hyperplane. The sharp-edge hyperplane is one that gives most prominent partition between the classes. There is dependably at least one support vector for each class and sometimes there are often more. In credit card counterfeit recognition, for each test data instance it decides whether the test example falls inside the learned area. At that point, if a test instance occurs inside the training area, then it is confirmed as expected otherwise anomalous. This model shows that it has a higher accuracy of detection when compared with other algorithms. It likewise has a superior time effectiveness and generalization capacity.

14.4 DATASET USED FOR THE MODEL

The most important requisite for performing the classification technique is the optimized use of dataset. Both the training and testing data of a classification model can be affected by the dimension of the dataset. For this study purpose, credit card fraud classification data (https://archive.ics.uci.edu/ml/machine-learningdatabases/00350/) have been used. The dataset has a total 690,000 instances in the dimension of 30,000 rows and 23 columns. For our study, 80% of the credit card data are considered for training purpose and similarly 20% of the credit card data for testing purpose because of its optimized value of accuracy in this ratio. The use of credit card fraud classification dataset can improve the efficiency of our research by saving the data collection time and data access time.

14.5 RESULT AND DISCUSSION

14.5.1 Experimental setup

Five classification algorithms are implemented with the help of MATLAB platform, version R2019a. During processing, an i7 processor with 3.4 GHz clock speed was used as the system. The main memory space and the secondary memory space in each system were 1 TB and 10 GB, respectively.

14.5.2 Evaluation of performance parameters

Confusion matrix: Confusion matrix is a way to represent positive and negative classes for the processing of classification models. It helps to demonstrate the total number of accurately and inaccurately classified instances that are contrasted with the true results (target value) in the testing data. A confusion matrix is a 2×2 table framework of binary class classification designed to sum the quantity of the results of four boxes of a binary classifier, and we more often denote them as true positive (TP), true negative (TN), false positive (FP), and false negative (FN).

A binary classification model predicts the test result of a dataset as either a positive class or a negative class for the data instances. This classification produces four results, for example, TP, FP, TN, and FN.

- TP class: A positive class is correctly predicted.
- FP class: A positive class is predicted incorrectly.
- TN class: A negative class is correctly predicted.
- FN class: A negative class is predicted incorrectly.

Various performance parameters have been evaluated from the confusion matrix for our proposed model.

14.5.2.1 Accuracy

Accuracy parameter is being evaluated to know that up to what extent the classifier is correct. It is calculated by dividing total number of correctly predicted classes with the total number of positive and negative classes. The best performed accuracy is assumed to be 1.0, and 0.0 is the worst. It can also be calculated by using the formula:

$$1 - \text{Error rate}. \tag{14.1}$$

The accuracy for all the five classification algorithms is calculated by using the confusing matrix.

$$\text{Accuracy} = \frac{(\text{TP} + \text{TN})}{(\text{TP} + \text{TN} + \text{FP} + \text{FN})}. \tag{14.2}$$

14.5.2.2 Precision

Precision helps to find the total positively classified relevant instances from the retrieved instances. We calculate precision by dividing the total number of correctly predicted classes with the total number of positive predictions. It is also described as positive predictive value. The best precision is 1.0, and 0.0 is worst precision.

$$\text{Precision} = \frac{\text{TP}}{(\text{TP} + \text{FP})}. \tag{14.3}$$

14.5.2.3 Sensitivity

Sensitivity is calculated to correctly identify the true positive rate and to estimate the error. It specifies that how good is the test to detect the positive classes. Sensitivity is calculated by dividing the total number of correctly positive predicted classes with the total number of positive classes. It is also known as recall or true positive rate. The best sensitivity is 1.0, and 0.0 is the worst sensitivity.

$$\text{Sensitivity} = \frac{\text{TP}}{(\text{TP} + \text{FN})}. \tag{14.4}$$

14.5.2.4 Specificity

It is evaluated to identify how accurately the algorithm identifies the false alarms. We calculate specificity by dividing the total number of correctly negative predicted classes with the total number of negative classes. It is also known as true negative rate. The best specificity is 1.0, whereas 0.0 is the worst.

$$\text{Sensitivity} = \frac{\text{TN}}{(\text{TP} + \text{FP})}. \tag{14.5}$$

14.5.2.5 FI score

In two class classification, F-measure or F1 score is a measure used to find the testing data accuracy. It uses both the precision as well as recall of the test to calculate the score.

$$\text{F1 score} = \frac{2\text{TP}}{(2\text{TP} + \text{FP} + \text{FN})}. \tag{14.6}$$

14.5.3 Proposed model

Five machine learning models such as naïve Bayes, ELM, KNN, MLP, and SVM are considered for classification of the data instances to find prediction accuracy. Among them, SVM model has the maximum predictive accuracy value of 81.40%.

Better predictive result can be achieved when a model is more robust and efficient. By combining the classification models or by assembling machine learning classifiers the performance is improved as compared with a single learning model [13]. More prominently, it helps to reduce the variance and increases the predictive accuracy value. Three classification models, SVM, KNN, and MLP, are combined from heterogeneous families to build

a combined model or hybrid model for finding a better predictive result. These three classifiers are chosen because their individual accuracies are better compared with the other two classification models. A proposed block diagram is shown in Figure 14.1. The decision boundary of the hyperplane is optimized by stacking of classification models.

The total number of data instances are shown in Table 14.1 with the help of a confusion matrix. The input data are the testing data in the proposed model. In the proposed model, the predictive accuracy value is obtained as 82.42%, and it is better than the values of the individual models. The separation boundary values reduce the predictive error for the two-class classification method. According to the results shown in Table 14.2, it is observed that the highest accuracy is obtained for the proposed model. By using this approach, a better accuracy can be achieved compared with the individual classification models.

In Table 14.2, the performance parameters provided are accuracy, precision, sensitivity, specificity, and F1 score. The values are calculated for different learning models. As shown in Table 14.2, the proposed

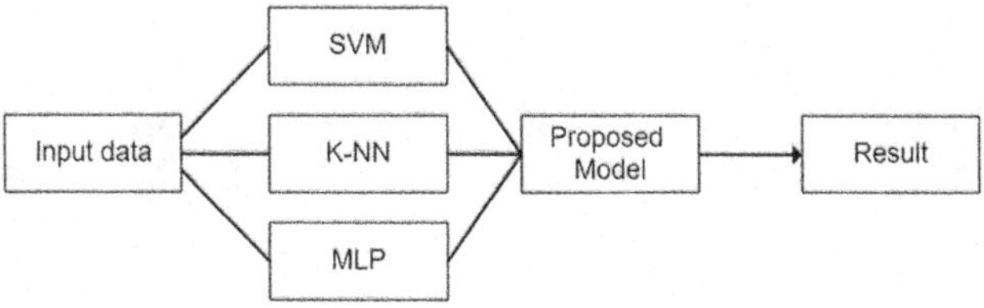

Figure 14.1 Block diagram for the proposed model.

Table 14.1 Confusion matrix for the proposed model

	Actual class	
Predicted class	True positive 4547	False positive 159
	False negative 896	True negative 398

Table 14.2 Accuracy results for various models

Classifiers parameters	*Naïve Bayes*	*ELM*	*KNN*	*MLP*	*SVM*	*Proposed model*
Accuracy	0.4990	0.7988	0.8040	0.8110	0.8140	0.8242
Precision	0.8875	0.8045	0.8218	0.9515	0.6937	0.9662
Sensitivity	0.4085	0.9799	0.9555	0.8444	0.2634	0.8354
Specificity	0.8176	0.1613	0.2705	0.6911	0.9670	0.7145
F1 score	0.5595	0.8836	0.8836	0.8948	0.3818	0.8960

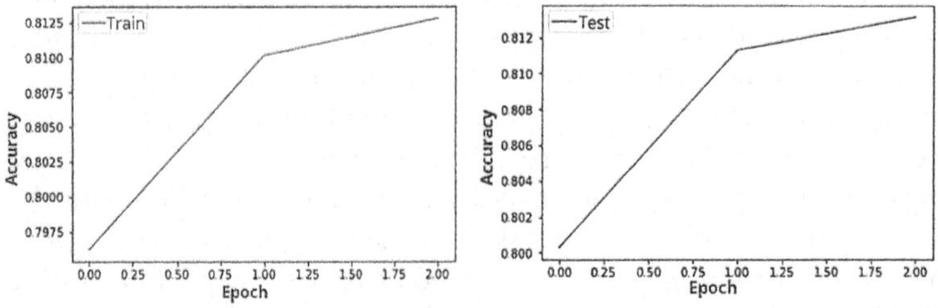

Figure 14.2 Accuracy values of proposed model for training and testing data.

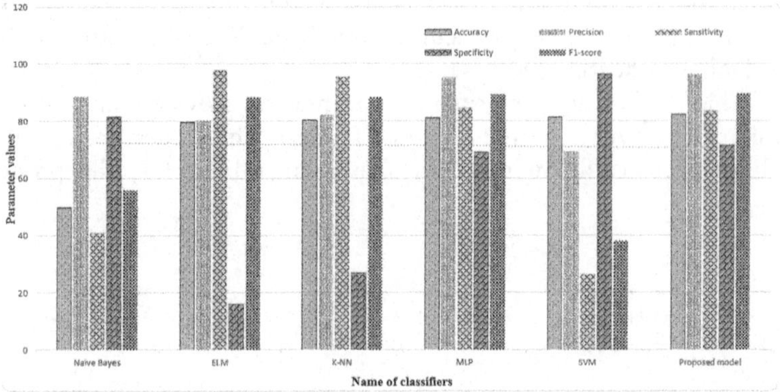

Figure 14.3 Comparison between classifiers for various parameters.

model yields better prediction accuracy as compared with other models. It is observed that the highest value of predictive accuracy turns out to be 82.42% for the proposed classification model. Among all the individual models, SVM model has the highest predictive accuracy.

In Figure 14.2, the values of accuracy for both training and testing data are shown. In the plotted graph, it can be observed that there is an increase in accuracy percentage with respect to periodical time.

In Figure 14.3, the comparison between various performance parameters like accuracy, precision, sensitivity, specificity, and F1 score is shown with respect to the number of classifiers we discussed as well as for the proposed model. It is observed that the predictive accuracy value is more for the proposed model as compared with other classification models.

14.6 CONCLUSION

Five important classification models of machine learning techniques are examined thoroughly in this study and their performance parameters are then compared along with the predictive accuracy between them.

While considering the prediction accuracy among the five classification models, the results reveal that there are marginal differences in the error rates of the five classification models. In the proposed model, the correct predictive class is observed to be more with minimized false alarms. The prediction accuracy was found to be 82.42% for the proposed model. The prediction accuracy for the proposed model is observed to be highest over the values obtained from individual values.

For the future work, we intend to apply various classification models from other machine learning families to improve the prediction accuracy along with other performance parameters.

REFERENCES

1. Dal Pozzolo, Andrea, Olivier Caelen, Yann-Ael Le Borgne, Serge Waterschoot, and Gianluca Bontempi. "Learned lessons in credit card fraud detection from a practitioner perspective." *Expert Systems with Applications*, vol. 41, no. 10, pp. 4915–4928, 2014.
2. Sahin, Yusuf, and Ekrem Duman. "Detecting credit card fraud by ANN and logistic regression." In *International Symposium on Innovations in Intelligent Systems and Applications (INISTA), 2011*, pp. 315–319. IEEE, 2011.
3. Kou, Yufeng, Chang-Tien Lu, Sirirat Sirwongwattana, and Yo-Ping Huang. "Survey of fraud detection techniques." In *IEEE International Conference on Networking, Sensing and Control, 2004*, vol. 2, pp. 749–754. IEEE, 2004.
4. Brause, R., T. Langsdorf, and Michael Hepp. "Neural data mining for credit card fraud detection." In *Proceedings of 11th IEEE International Conference on Tools with Artificial Intelligence, 1999*, pp. 103–106. IEEE, 1999.
5. Jagielska, Ilona, and Janusz Jaworski. "Neural network for predicting the performance of credit card accounts." *Computational Economics*, vol. 9, no. 1, pp. 77–82, 1996.
6. LNCS Dorronsoro, Jose R., Francisco Ginel, Carmen R. Sánchez, and Carlos Santa Cruz. "Neural fraud detection in credit card operations." *IEEE Transactions on Neural Networks*, vol. 8, pp. 827–834, 1997.
7. Bahnsen, Alejandro Correa, Djamila Aouada, Aleksandar Stojanovic, and Björn Ottersten. "Feature engineering strategies for credit card fraud detection." *Expert Systems with Applications*, vol. 51, pp. 134–142, 2016.
8. Ng, Andrew Y., and Michael I. Jordan. "On discriminative vs. generative classifiers: A comparison of logistic regression and Naive Bayes." In *Advances in Neural Information Processing Systems*, pp. 841–848. 2002.
9. Huang, Guang-Bin, Qin-Yu Zhu, and Chee-Kheong Siew. "Extreme learning machine: A new learning scheme of feedforward neural networks." In *Proceedings 2004 IEEE International Joint Conference on Neural Networks, 2004*, vol. 2, pp. 985–990. IEEE, 2004.
10. Fukunaga, Keinosuke, and Patrenahalli M. Narendra. "A branch and bound algorithm for computing k-nearest neighbors." *IEEE Transactions on Computers*, vol. 100, no. 7, pp. 750–753, 1975.

11. Collobert, Ronan, and Samy Bengio. "Links between perceptrons, MLPs and SVMs." In *Proceedings of the Twenty-First International Conference on Machine Learning*, p. 23. ACM, 2004.
12. Hearst, Marti A., Susan T. Dumais, Edgar Osuna, John Platt, and Bernhard Scholkopf. "Support vector machines." *IEEE Intelligent Systems and Their Applications*, vol. 13, no. 4, pp. 18–28, 1998.
13. Dietterich, Thomas G. "Ensemble methods in machine learning." In *International Workshop on Multiple Classifier Systems*, pp. 1–15. Springer, Berlin, Heidelberg, 2000.

Chapter 15

Authentication in RFID scheme based on elliptic curve cryptography

Abhay Kumar Agrahari and Shirshu Varma
Indian Institute of Information Technology Allahabad

CONTENTS

15.1 INTRODUCTION

Radio frequency identification (RFID) is a wireless communication device that means two different objects, which are placed in different places, communicate with each other using the radio waves. An RFID system uses three components for communication: tag, reader, and a backend server.

Following are the uses of these components.

RFID tag: It is a radio transponder composed of a circuit to store and process identification information. The RFID tag consists of an antenna which is used for communication with RFID readers [1,2].

RFID reader: It is the wireless device used to transmit and receive signals which come from RFID tag. It also acts as a path between the backend server and tag for communication.

Backend server: Backend server is a database which stores the data of RFID tags. The stored data are primarily used for the authentication of RFID tags.

In RFID, a channel is wireless and insecure between RFID tag and RFID reader. Whereas RFID reader and backend server are connected with wired channel, so it is a secure channel. So, we will secure this wireless channel, and to authenticate tags use many authentication protocols. The communication between the RFID tag and RFID reader is along a wireless channel and is therefore insecure, in contrast to the communication between the RFID reader and the backend server, which takes place along a secured connected channel. The aim is to secure this wireless channel and authenticate tags using authentication protocols. There are many types of authentication protocols which use concepts like symmetric key cryptography, public key cryptography or hash function. In many authentication protocols, symmetric key cryptography is used [3–8]. The reason behind using symmetric key cryptography is its faster computation compared with public key cryptography. But symmetric-key-based solutions face scalability problem as the backend server has to use a linear search for tag identification. Because of this, the use of public key cryptography for authenticating protocols came into existence. Elliptic curve cryptography (ECC)–based RFID authentication protocols make use of public key cryptography.

In 2006, the authentication protocol for RFID using ECC was proposed by Tuyls and Batina [9], which was based on identification schnorr scheme. In the next year (2007), using the Okamoto identification scheme, Batina et al. [10] implemented an ECC-based authentication protocol. But in 2008, Lee et al. [11] proved that a privacy problem exists in both Tuyls–Batina's and Batina's protocols. To solve this problem, Lee et al., Godor et al. [12], and O'Neill and Robshaw [13] implemented an improved RFID authentication protocol which used ECC. But these three protocols suffered from scalability problem which is discussed in Chou's [14] authentication protocol for RFID using ECC.

In this paper, we discuss that Chou's protocol suffered from some authentication-based attack as well as tag cloning attack and impersonation attack. Our protocol is an improved version of Chou's protocol. The improved ECC-based authentication protocols are used in many places like internet of things, healthcare, etc. Some improved algorithms which used ECC and also worked in real-time environments were also proposed [15,16].

This paper is organized as follows: Section 15.2 describes why the elliptic curve is used for RFID authentication protocol, Section 15.3 covers the working of the Chou's protocol. Section 15.4 discusses the frailty of Chou's protocol, and Section 15.5 introduces an enhanced and improved protocol for RFID authentication. Lastly, Section 15.6 provides a comparison of security requirements and computational costs.

15.2 ELLIPTIC CURVE CRYPTOGRAPHY REVIEW

In elliptic curve (E), the points on the X-Y plane satisfy the Weierstrass equation, which is applied over a finite field of the form F_q.

ECC is used due to its small key size for the same level of security. Using public key cryptography, many RFID protocols [17–22,23,24] have been implemented for authentication purpose in order to improve the security in RFID.

15.2.1 Elliptic curve discrete logarithm problem

Given any elliptic curve E, over the finite field F_q it is expressed as $E(F_q)$. There is a point P which belongs to that finite field with order n and a point $Q \in \langle P \rangle$. There exists a hard problem to find an integer $\alpha \in [0, n-1]$ from given P and Q, where $Q = aP$ [25].

15.2.2 Elliptic curve factorization problem

Given two points P and Q over a finite field, where $R = xP + yQ$, it is hard to find two points xP and yQ over the finite field.

15.2.3 Diffie–Hellman problem in elliptic curve

From given three points P, xP, yQ, it is hard to find xyP over the finite field.

15.3 REVIEW OF CHOU'S PROTOCOL

In this section, we review the Chou's protocol which is used for authentication. RFID Chou's authentication protocol has two different phases: a setup phase and an authentication phase.

Some symbols used in RFID Chou's protocol are defined as follows:

- q a predefined large prime number,
- G a group with order n having additive properties consisting of points on an elliptic curve,
- P a group (G) generator,
- X_i the value of the ith tag identifier which is a random point in G,
- y the server's private key,
- Y the server's public key, where $Y = yP$,
- r, k two random numbers,
- $h(\ )$ a secure hash function.

15.3.1 Chou's protocol setup phase

The setup phase consists of a server that generates two keys, public and private, and also generates an identifier for each tag.

Setup phase has two phases:

In the first phase, server generates a number randomly from $y \in Z_q$ and uses it as the private key, then calculates Y from $Y{=}yP$ and uses it as the public key.

In the second phase, the ith tag identifier chooses randomly from "G", that is, X_i. The value X_i is then saved in a server database. In the tag's memory, the server stores value $[X_i,\ Y,\ P]$ for future use.

15.3.2 Chou's protocol authentication phase

Server y, Y, X_i	Tag X_i, Y, P
Generate $r = Z_q$	Generate $k = Z_q$
$C_0 = rY$	$K = kP$
$\rightarrow C_0$	$C_1 = kC_0$
$K = y^{-1}r^{-1}C_1$	$R = K + K$
$R = K + K$	$C_2 = X_i + R$
$X_i = C_2 - R$	$C_3 = h(X_i,\ K)$
Verify $C_3 = h\ (X_i,\ K)$	$\leftarrow C_1, C_2, C_3$
Search X_i	Verify $C_4 = h(X_i,\ 3K)$
$C_4 = h(X_i,\ 3K)$	
$\rightarrow C_4$	

15.4 PROBLEMS IN CHOU'S AUTHENTICATION PROTOCOL

In Chou's RFID authentication protocol, there are some security problems. According to Chou, his protocol is secure against many attacks which occur when authenticating a server and a tag.

15.4.1 Problem in tag privacy

Tag privacy means that attacker is unable to get the tag's identity X_i. In Chou's protocol, it is easy to find the tag identity without any intention of attack, if only the value of P is known. Let us assume that$A$ is an attacker, then it can easily find the tag identifier using the following steps:

1. The attacker A acts like a server and generates a random number $r \in Z_q$. Then it sends $C_0 = r$ to the tag.
2. After receiving this message from the attacker, the tag, considering this message came from the server, computes

   Generate $k \in Z_q$

   $$K = kP,$$

   $$C_1 = kC_0,$$

   $$R = K + K,$$

   $$C_2 = X_i + R,$$

   $$C_3 = h(X_i, K).$$

3. The attacker intercepts the message and computes the tag identity as follows:

   $$C_2 - 2r^{-1}.C_1.P,$$

   $$X_i + 2K - 2r^{-1}.k.C_0.P,$$

   $$X_i + 2K - 2r^{-1}.k.\mathrm{r}.P,$$

   $$X_i + 2K - 2K,$$

   $$X_i.$$

15.4.2 Problem in mutual authentication

When the attacker gets the value C_1, it computes the tag's identity X_i using the equations as described earlier. Since the value of K and X_i are known to the attacker, it continues the attack described previously and sends $C_4 = h(X_i, 3K)$to the tag. After getting C_4 from the attacker, the tag compares this value to $h(X_i, 3K)$ which is equal. So, tag knows that it has come from a legal server. This is referred to as the mutual authentication problem which occurs in Chou's protocol.

15.4.3 Forward traceability problem

Let us assume that the attacker could get the tag's identity X_i, then it could also confirm whether the message is coming from tag or not by tracing that particular tag.

1. Let the attacker get the tag's identifier X_i.
2. The attacker collects the message $(C_0, C_1, C_2, C_3, C_4)$ transmitted between the tag and server, where $C_0 = r.Y$, $K = k.P$, $C_1 = k.C_0$, $R = K + K$, $C_2 = X_i + R$, $C_3 = h(X_i, K)$ and $C_4 = h(X_i, 3K)$.
3. The attacker then computes $R' = C_2 - X_i^{'}$, $K' = 2^{-1}.R'$ and verifies whether C_3 and $h(X_i^{'}, K')$ are equal. If both are equal, then the attacker confirms that message came from tag.

Because of the aforementioned attacks, we can verify that Chou's protocol cannot withstand the tag privacy problem, mutual authentication problem, and "forward traceability problem. So, a new enhanced protocol has been proposed to solve these problems.

15.5 PROPOSED PROTOCOL

The new enhanced protocol is used to overcome the problems that exist in Chou's authentication protocol. This protocol also has two phases, the first phase is the same as the setup phase in Chou's protocol, while the second phase is a new authentication phase.

15.5.1 Setup phase

Similar to the Chou's protocol setup phase, the server generates two keys, public and private, and also generates an identifier for each tag.

Setup phase has two phases:

In the first phase, the server generates a number randomly from $y \in Z_q$ and uses it as the private key, it then calculates Y from $Y=yP$ and uses it as the public key.

In the second phase, the ith tag identifier chooses randomly from "G", that is, X_i. The value X_i is then saved in a server database. In the tag's memory, the server stores value $[X_i, Y, P]$ for future use.

15.5.2 New authentication phase

In new authentication protocol, there are various transitions between server and tag. They both (server and tag) will authenticate each other, the steps are as follows:

1. Server will generate a random number r, such that $r \in Z_q$ and then calculate $C_0 = r$. It then sends these data to the tag for further authentication process.
2. The tag on receiving C_0 from the server generates a random number k, from $k \in Z_q$ and then computes

   $K = kY$, and

   $C_1 = kP$.

   The tag then encrypts C_0, C_1, K using the server's public key Y and adds it to the tag's identity X_i, and computes

   $C_2 = X_i + \text{Enc}_Y(C_0, C_1, K)$.

The value of C_1 and C_2 are sent to the server.

3. After receiving C_1 and C_2 from the tag, the server will verify the tag. First, it will find K' from $K' = yC_1$ and the tag's identity from $X_i = C_2 - \text{Dec}_y(C_0, C_1, K')$ to get X_i'', the server decrypts the data using its private key (y) and after that subtracts it from C_2 to get X_i'. It then finds X_i'' in the database. If it exists, the server will authenticate the tag, otherwise it will terminate the session.

   Server will generate a hash value $C_3 = h(X_i', K')$ and send it to the tag.
4. In this step, tag will verify that server is a legal server or not. To check this condition, the tag will generate a hash value $h(X_i, K)$. If this hash value is equal to C_3, then tag believes that the server is a legal server.

Server y, Y, X_i	Tag X_i, Y, P
Generate $r = Z_q$	Generate $k = Z_q$
$C_0 = r$	$K = kY$
$\rightarrow \boldsymbol{C_0}$	$C_1 = kP$
$K' = yC_1$	$C_2 = X_i + \text{Enc}_Y(C_0, C_1, K)$
$X_i' = C_2 - \text{Dec}_y(C_0, C_1, K')$	$\leftarrow \boldsymbol{C_1, C_2}$
Search X_i'	Verify $C_3 = h(X_i, K)$
$C_3 = h(X_i', K')$	
$\rightarrow \boldsymbol{C_3}$	

15.6 SECURITY ANALYSIS OF PROPOSED PROTOCOL

Here, we are exploring the proposed ECC-based RFID authentication protocol for security analysis. We can signify that the proposed RFID authentication protocol is secure against tag privacy problem, forward traceability problem, and mutual authentication problem, and it can also withstand some basic attacks (tag impersonation attack, replay attack, etc.).

15.6.1 Secure against tag privacy attack

In Chou's algorithm, when attacker knows the value of "P", then it can easily find the tag's identity X_i. Let us suppose that the attacker will generate a random number r from $r \in Z_q$ and compute $C_0 = r$ and send a message C_0 to the tag. Then after getting this message from server, tag will randomly choose a number k from $k \in Z_q$ and calculate

$$K = kY,$$

$$C_1 = kP,$$

$$C_2 = X_i + \mathrm{Enc}_Y(C_0, C_1, K).$$

Then tag will send the message C_1, C_2 to the attacker. So, if attacker wants to know the tag's identity X_i from C_2, it has to compute $K = kY$ from $C_1 = kP$ and $Y = yP$. Here, the attacker will face a Diffie–Hellman problem as well as should know the private key (y) of server for decryption $\mathrm{Dec}_y(C_0, C_1, K)$. So, the proposed RFID authentication protocol will overcome the problem of tag privacy.

15.6.2 Secure against mutual authentication problem

Without knowing the tag's identifier X_i and K, the attacker is unable to generate a message $C_3 = h(X_i, K)$. So, the attacker is unable to send these data to the tag and the tag will be unable to verify the correctness of C_3, that is, unable to verify from where did it come from. So, the proposed RFID authentication protocol is secure against mutual authentication problem.

15.6.3 Secure against forward traceability problem

Let us assume that the attacker could get that tag's identity d and also intercept the message between server and tag, that is, (C_0, C_1, C_2, C_3), where $C_0 = r$, $C_1 = kP$, $C_2 = X_i + \mathrm{Enc}_Y(C_0, C_1, K), C_3 = h(X_i', K')$. But to verify whether those messages are transmitted between server or tag, it has to calculate the value K from $K = kY$, $C_1 = kP$, and $Y = yP$. So, here the attacker

will face a Diffie–Hellman problem. Thus, the new RFID authentication protocol is secure against the traceability problem.

15.6.4 Secure against replay attack

In replay attack, the attacker decodes the message C_0 and replays it to the tag. When tag receives this message, it can generate a new message C_1, C_2 and send it to the attacker. But after getting this message from tag, attacker cannot generate C_3 without the knowledge of tag's identifier X_i and value of K because the attacker has no knowledge of server's private key. So, the tag cannot verify the replay attack by checking the correctness of C_3.

15.6.5 Secure against tag impersonation attack

Suppose the attacker impersonates the *i*th tag to the server. When the attacker intercepts the message C_0 from the server, it generates a message (C_1, C_2) where $K = kY$, $C_1 = kP$, $C_2 = X_i + \mathrm{Enc}_Y(C_0, C_1, K)$. However, the attacker is unable to generate a legal message C_2 without the tag's identifier. Thus, the proposed protocol is free from tag impersonation attack.

15.6.6 Secure against modification attack

Suppose the attacker intercepts the message C_0 or C_3, and sends it to the tag after modification. In the new protocol, the tag can identify the attack by checking the correctness of C_3. If the attacker intercepts the message (C_1, C_2) and sends to the server after modification, the server can also identify the attack by verifying X_i from the database. So, the new protocol is secure against modification attack.

15.7 PERFORMANCE ANALYSIS OF PROPOSED PROTOCOL

In this section, we compare our proposed ECC-based RFID authentication protocol to some existing authentication protocols in terms of security as well as computational cost. As it is well known that tag computing capability and memory are very limited, computational costs and security requirements are important characteristics for practical applications. An elliptic curve defined over the finite field of the form $F(2^{163})$ is used in many applications.

15.7.1 Analysis of security requirements of ECC-based protocol

In this section, we compare some other ECC-based RFID authentication protocols with our proposed protocol. In Table 15.1, we compare the security requirements.

Table 15.1 Comparison in security requirements

Attacks	*Tuyls's*	*Batina's*	*Lee's*	*Chou's*	*Ours*
Replay attack	Secure	Secure	Secure	Secure	Secure
Man in middle attack	Not secure	Not secure	Not secure	Secure	Secure
Impersonation attack	Not secure	Not secure	Not secure	Not secure	Secure
Mutual authentication	Not secure	Not secure	Not secure	Not secure	Secure
Tag privacy	Not secure	Not secure	Secure	Not secure	Secure
Forward traceability	Not secure	Not secure	Secure	Not secure	Secure

15.7.2 Analysis of computational cost

Let us assume

CT_{mul} be the computational time for multiplication operation,
CT_{inv} be the computational time for inversion operation,
CT_{eca} be the computational time for elliptic curve point addition operation,
CT_{ecm} be the computational time for elliptic curve point multiplication operation, and
CT_h be the computational time for hash operation.

According to Chatterjee et al.'s work [26], we have $CT_{inv} \approx 3.CT_{mul}$, $CT_{eca} \approx 5.CT_{mul}$, $CT_{ecm} \approx 1{,}200.CT_{mul}$, $CT_h \approx .36\ .CT_{mul}$. The detailed comparison of existing ECC-based RFID authentication protocols with our proposed protocols is done in Table 15.2. In Table 15.2, we compare computational costs. Here, we compare that our proposed protocol using the concept of hash function and encryption takes less time as compared with other algorithms, like Lee's EC RAC-1, RAC-2, RAC-3 algorithms that use the concept of elliptic curve to implement the secure authentication RFID protocol. But EC RAC-3 is more complex as compared with EC RAC-1 and EC RAC-2, and EC RAC-2 is more complex as compared with EC RAC-1. So different algorithms use different concepts to create a secure authenticated RFID protocol.

15.8 CONCLUSION

The ECC-based RFID authentication protocol is the most secure protocol for providing security service for internet of things. Here we examine Chou's protocol and describe the weaknesses of this protocol, for example, Chou's protocol is not secure against mutual authentication, tag privacy problem, and forward traceability problem. In our work, first the possible attacks in Chou's protocol have been discussed. Chou's protocol is vulnerable to attacks as the attacker can break the authentication by sending the value of $C_0 = r$ and obtaining the tag's identity. In the proposed ECC-based authentication protocol, we can change the value of K, C_0 and also use the

Table 15.2 Comparison in computational cost

Scheme	*Tag side computational cost*	*Server side computational cost*	*Total cost*
Lee et al.'s EC RAC-I scheme [27]	$1\times CT_{mul}+2\times CT_{ecm}\approx 2{,}401\times CT_{mul}$	$2\times CT_{inv}+1\times CT_{eca}$ $+2\times CT_{ecm}\approx 2{,}411\times CT_{mul}$	$4{,}812\times CT_{mul}$
Lee et al.'s EC RAC-2 scheme [27]	$3\times CT_{mul}+3\times CT_{ecm}\approx 3{,}603\times CT_{mul}$	$4\times CT_{inv}+2\times CT_{eca}+5\times CT_{ecm}$ $\approx 6{,}022\times CT_{mul}$	$9{,}605\times CT_{mul}$
Lee et al.'s EC RAC-3 scheme [27]	$3\times CT_{mul}+4\times CT_{ecm}\approx 4{,}803\times CT_{mul}$	$4\times CT_{inv}+2\times CT_{eca}+5\times CT_{ecm}$ $\approx 6{,}022\times CT_{mul}$	$10{,}825\times CT_{mul}$
Godar and Imre's scheme [28]	$2\times CT_{mul}+5\times CT_{ecm}\approx 6{,}022\times CT_{mul}$	$2\times CT_{mul}+5\times CT_{ecm}$ $\approx 6{,}022\times CT_{mul}$	$12{,}004\times CT_{mul}$
Chen et al.'s scheme [29]	$2\times CT_{mul}+4\times CT_{ecm}+2\times CT_{inv}+1\times CT_{eca}$ $+5\times CT_{h}\approx 4{,}815\times CT_{mul}$	$2\times CT_{ecm}+2\times CT_{inv}+1\times CT_{eca}$ $+3\times CT_{h}\approx 2{,}412\times CT_{mul}$	$7{,}227\times CT_{mul}$
Batina et al.'s scheme [30]	$1\times CT_{mul}+3\times CT_{ecm}\approx 3{,}601\times CT_{mul}$	$2\times CT_{inv}+7\times CT_{ecm}$ $+3\times CT_{eca}\approx 8{,}421\times CT_{mul}$	$12{,}022\times CT_{mul}$
Liu et al.'s scheme [31]	$2\times CT_{inv}+1\times CT_{ecm}+3\times CT_{mul}\approx 1{,}212\times CT_{mul}$	$2\times CT_{ecm}+2\times CT_{mul}$ $\approx 2{,}402\times CT_{mul}$	$3{,}614\times CT_{mul}$
Wang et al.'s scheme [32]	$1\times CT_{b}+2\times CT_{ecm}+2\times CT_{mul}\approx 2{,}402\times CT_{mul}$	$1\times CT_{eca}+2\times CT_{ecm}$ $+2\times CT_{b}\approx 2{,}402\times CT_{mul}$	$4{,}804\times CT_{mul}$
Chou's scheme	$3\times CT_{mul}+2\times CT_{ecm}+2\times CT_{b}\approx 2{,}404\times CT_{mul}$	$1\times CT_{mul}+2\times CT_{ecm}+2\times CT_{b}$ $+2\times CT_{inv}\approx 2{,}408\times CT_{mul}$	$4{,}812\times CT_{mul}$
Proposed protocol	$1\times CT_{eca}+2\times CT_{ecm}+1\times CT_{b}\approx 2{,}406\times CT_{mul}$	$1\times CT_{eca}+1\times CT_{ecm}$ $+1\times CT_{b}\approx 1{,}206\times CT_{mul}$	$3{,}612\times CT_{mul}$

encryption using the server's public key to secure our protocol, so our proposed protocol is more secure as compared with other existing protocols. We compare computational cost on the bases of how much computation occurs between the server and tag, on the tag side, as well as server side. At last, we conclude that the proposed protocol is much secure and less costly as compared with existing protocols.

REFERENCES

1. Juels, A., and Weis, S. A. 2009. Defining strong privacy for RFID. *ACM Transactions on Information and System Security (TISSEC)*. 13(1): 7.
2. Cai, S., Li, Y., Li, T., and Deng, R. H. 2009. Attacks and improvements to an RIFD mutual authentication protocol and its extensions. In *Proceedings of the Second ACM Conference on Wireless Network Security* (pp. 51–58). ACM.
3. Dehkordi, M. H., and Farzaneh, Y. 2014. Improvement of the hash-based RFID mutual authentication protocol. *Wireless Personal Communications*. 75(1): 219–232.
4. Safkhani, M., Peris-Lopez, P., Hernandez-Castro, J. C., and Bagheri, N. 2014. Cryptanalysis of the Cho et al. protocol: A hash-based RFID tag mutual authentication protocol. *Journal of Computational and Applied Mathematics*. 259: 571–577.
5. Alagheband, M. R., and Aref, M. R. 2014. Simulation-based traceability analysis of RFID authentication protocols. *Wireless Personal Communications*. 77(2): 1019–1038.
6. Chen, C. L., Huang, Y. C., and Shih, T. F. 2012. Novel mutual authentication scheme for RFID conforming EPCglobal class 1 generation 2 standards. *Information Technology and Control*. 41(3): 220–228.
7. Kuo, W. C., Chen, B. L., and Wuu, L. C. 2013. Secure indefinite-index RFID authentication scheme with challenge-response strategy. *Information Technology and Control*. 42(2): 124–130.
8. Alagheband, M. R., and Aref, M. R. 2013. Unified privacy analysis of newfound RFID authentication protocols. *Security and Communication Networks*. 6(8): 999–1009.
9. Tuyls, P., and Batina, L. 2006. RFID-tags for anti-counterfeiting. In *Cryptographers Track at the RSA Conference* (pp. 115–131). Springer, Berlin, Heidelberg.
10. Batina, L., Guajardo, J., Kerins, T., Mentens, N., Tuyls, P., and Verbauwhede, I. 2007, March. Public-key cryptography for RFID-tags. In *Null* (pp. 217–222). IEEE.
11. Lee, Y. K., Batina, L., and Verbauwhede, I. 2008. EC-RAC (ECDLP based randomized access control provably secure RFID authentication protocol. In *IEEE International Conference on RFID, 2008* (pp. 97–104). IEEE.
12. Gdor, G., Giczi, N., and Imre, S. 2010. Elliptic curve cryptography based mutual authentication protocol for low computational capacity RFID systems performance analysis by simulations. In *Networking and Information Security (WCNIS), 2010 IEEE International Conference on Wireless Communications* (pp. 650–657). IEEE.

13. O'Neill, M., and Robshaw, M. J. 2010. Low-cost digital signature architecture suitable for radio frequency identification tags. *IET Computers & Digital Techniques*. 4(1): 14–26.
14. Chou, J. S. 2014. An efficient mutual authentication RFID scheme based on elliptic curve cryptography. *The Journal of Supercomputing*. 70(1): 75–94.
15. Jin, C., Xu, C., Zhang, X., and Li, F. 2016. A secure ECC-based RFID mutual authentication protocol to enhance patient medication safety. *Journal of Medical Systems*. 40(1): 12.
16. Alamr, A. A., Kausar, F., Kim, J., and Seo, C. 2018. A secure ECC-based RFID mutual authentication protocol for internet of things. *The Journal of Supercomputing*. 74(9): 4281–4294.
17. Burmester, M., De Medeiros, B., and Motta, R. 2008. Robust, anonymous RFID authentication with constant key-lookup. In *Proceedings of the 2008 ACM Symposium on Information, Computer and Communications Security* (pp. 283–291). ACM.
18. Gaubatz, G., Kaps, J. P., Ozturk, E., and Sunar, B. 2005. State of the art in ultra-low power public key cryptography for wireless sensor networks. In *Third IEEE International Conference on Pervasive Computing and Communications Workshops, 2005. PerCom 2005 Workshops* (pp. 146–150). IEEE.
19. Kaya, S. V., Sava, E., Levi, A., and Eretin, O. 2009. Public key cryptography based privacy preserving multi-context RFID infrastructure. *Ad Hoc Networks*. 7(1): 136–152.
20. Furbass, F., and Wolkerstorfer, J. 2007. ECC processor with low die size for RFID applications. In *IEEE International Symposium on Circuits and Systems, 2007. ISCAS 2007* (pp. 1835–1838). IEEE.
21. Lee, Y. K., Sakiyama, K., Batina, L., and Verbauwhede, I. 2008. Elliptic-curve-based security processor for RFID. *IEEE Transactions on Computers*. 57(11): 1514–1527.
22. Hutter, M., Feldhofer, M., and Plos, T. 2010. An ECDSA processor for RFID authentication. In *International Workshop on Radio Frequency Identification: Security and Privacy Issues* (pp. 189–202). Springer, Berlin, Heidelberg.
23. Qian, Q., Jia, Y. L., and Zhang, R. 2016. A lightweight RFID security protocol based on elliptic curve cryptography. *IJ Network Security*. 18(2): 354–361.
24. Benssalah, M., Djeddou, M., and Drouiche, K. 2016. Design and implementation of a new active RFID authentication protocol based on elliptic curve encryption. In *SAI Computing Conference (SAI)* (pp. 1076–1081). IEEE.
25. Hankerson, D., Menezes, A., and Vanstone, S. 2003. *Guide to Elliptic Curve Cryptography*. Springer - Verlag, New York.
26. Chatterjee, S., Das, A. K., and Sing, J. K. 2014. An enhanced access control scheme in wireless sensor networks. *Adhoc & Sensor Wireless Networks*. 21(1), 121–149.
27. Lee, Y. K., Batina, L., and Verbauwhede, I. 2010. Privacy challenges in RFID systems. In *The Internet of Things* (pp. 397–407). Springer, New York, NY.
28. Gdor, G., and Imre, S. 2011. Elliptic curve cryptography based authentication protocol for low-cost RFID tags. In *IEEE International Conference on RFID-Technologies and Applications (RFID-TA), 2011* (pp. 386–393). IEEE.

29. Chen, Y., Chou, J. S., Lin, C. F., and Wu, C. L. 2011. A novel RFID authentication protocol based on elliptic curve cryptosystem. *IACR Cryptology ePrint Archive*, 381.
30. Batina, L., Seys, S., Singele, D., and Verbauwhede, I. 2011. Hierarchical ECC-based RFID authentication protocol. In *International Workshop on Radio Frequency Identification: Security and Privacy Issues* (pp. 183–201). Springer, Berlin, Heidelberg.
31. Liu, Y. L., Qin, X. L., Wang, C., and Li, B. H. 2013. A lightweight RFID authentication protocol based on elliptic curve cryptography. *Journal of Computers*, 8(11).
32. Wang, S., Liu, S., and Chen, D. 2012. Analysis and construction of efficient RFID authentication protocol with backward privacy. In *China Conference on Wireless Sensor Networks* (pp. 458–466). Springer, Berlin, Heidelberg.

Chapter 16

Iris-based privacy-preserving biometric authentication using NTRU homomorphic encryption

E. Devi and P. P. Deepthi

National Institute of Technology Calicut

CONTENTS

16.1 INTRODUCTION

In the current digital world, several applications need identity verification of clients. Traditional methods for identity verification are based on either knowledge of the user (such as user ID and password) or tokens. These methods suffer from the disadvantage that password can be hacked and tokens can be stolen. In a biometric authentication system, biometric traits are used to verify the identity of the client. The biometric characteristics include face, iris, fingerprint, retina, and DNA. These biometric traits are unique to an individual facilitating implementation of accurate and convenient biometric authentication system, since there is no need for the user to carry tokens or remember a password. Among all these biometric traits, iris has gained popularity due to its robustness. Biometric systems have two phases of operation: enrolment and authentication (either identification or verification). Enrolment is the registration phase, in which the biometric trait is captured and the extracted feature templates are stored in

a database. In the authentication phase, fresh biometric template is compared with the registered templates in a database either to give access or to deny. Due to the digital revolution, most of the devices are using the services of the internet and storing the data in the cloud. Biometric traits are very sensitive and once leaked, they can neither be replaced nor revoked. Therefore, if the biometric template is stored in the understandable form, it will cause privacy and security issues. A user might want to prevent theft of his biometric, while a service provider wants to prevent the user from learning anything about the database. To address the security and privacy challenges, authentication and encryption schemes have been developed. It is desirable to develop encryption scheme to protect the biometric template, ensuring privacy of a user thereby providing trust between client and server. But the encryption process should be developed to support authentication in encrypted domain. All biometric authentication methods are threshold methods since features presented for verification will not be the same as the registered ones. This paper aims to design and implement privacy-preserving biometric authentication system.

16.2 RELATED WORK

In the literature, there are three basic approaches to privacy-preserving biometric authentication:

- Feature transformation includes cancellable biometrics [1,2] and bio-hashing [3]. These methods are adapted to real-time applications, and these are not secure if client-specific key is compromised.
- Biometric cryptosystems, based on error-correcting codes, include fuzzy commitment [4] and fuzzy vault [5], and these are neither practical nor secure.
- Homomorphic encryption (HE) [6], in which HE is used to protect the feature data and secure domain classification, is done from encrypted data.

HE is an algorithm which allows computations on ciphertext of a message to map computations on plaintext. The two group operations that are preserved are the arithmetic addition and multiplication.

- A HE is additive, if $E(a) \otimes E(b) = E(a + b)$,
- A HE is multiplicative, if $E(a) \otimes E(b) = E(a.\ b)$,

where E is encryption function and $\otimes$ the operation depending on user cipher and plaintext messages.

Three types of homomorphic cryptosystems are [7] partial homomorphic encryption (PHE), fully homomorphic encryption (FHE), and somewhat

homomorphic encryption (SHE). PHE allows either addition or multiplication on the ciphertext. SHE allows any one operation an arbitrary number of times and another operation limited number of times on ciphertext. FHE allows both the operations arbitrary number of times on ciphertext. Even though FHE [8,9] allows both the operations, PHE algorithms are better in terms of computational complexity and accuracy. HE allows an encrypted database to be stored in an untrusted third-party cloud where the processing operations can be performed without revealing the contents of the database. The first HE scheme is RSA (*N*-th degree Truncated polynomial Ring Units), which is a multiplicative homomorphic scheme. The traditional public key cryptosystems like RSA and ElGamal are partial homomorphic schemes. But these conventional methods of encryption schemes can be broken with quantum algorithms. Post-quantum cryptosystems are the replacement for traditional methods, and these are immune to quantum computers. Lattices have several hard mathematical problems which are not solvable by quantum computers, like shortest and closest lattice vector problems which give new era of a cryptosystem known as post-quantum cryptography. Lattice-based encryption and exploiting the additive homomorphic property of NTRU [10] give robust and secure biometric authentication and matching.

Iris is a popular biometric trait due to the property that it is very robust. It is highly unique due to which chances of having the same iris pattern for different persons is minimal; also, its epigenetic features remain unchanged over a lifetime. Literature survey says that variability of iris patterns among different persons is enormous and also it is tough to modify the iris pattern by any surgery. Hence iris could be a suitable choice for implementing an accurate and secure biometric authentication system.

16.3 FEATURE EXTRACTION FROM IRIS IMAGE

Feature extraction is the fundamental step of biometric authentication. The complete accuracy of a system depends on this step. Daugman [11,12] implemented an efficient iris-based authentication. The most discriminating feature of an iris is the phase information. Daugman's method is used to extract this phase information from an eye iris.

16.3.1 Iris segmentation

The first step in the feature extraction module is preprocessing. If the database is noise free, images can be directly taken to segmentation stage. The process of isolating the iris region from eye image is known as iris segmentation. The input of this stage is an eye image, and the output is an eye with circular boundaries around iris/sclera and pupil/iris. The total accuracy of an iris authentication system depends on the segmentation of iris region

from an eye image. The Daugman's integro differential operator defined by Eq. (16.1) is used for segmentation process.

$$\text{Max}_{(r,x_0,y_0)}\left|G\sigma(r)\frac{\partial}{\partial r}\oint_{(r,x_0,y_0)}\frac{I(x,y)}{2\pi r}\partial s\right|, \tag{16.1}$$

where $I(x, y)$ is eye image; $G\sigma(r)$, a Gaussian smoothing function; r, radius to search; and s, the contour of a circle given by r, x_0, y_0. The eyelids are removed by linear Hough transform, and a simple thresholding method is used to remove eyelashes because the eyelashes are quite dark as compared with rest of the eye.

16.3.2 Iris normalization

The next step of feature extraction is normalization. Normalization is the process of transforming concentric iris region to non-concentric form to get a constant dimension of a template to allow comparisons. Figure 16.1 describes the Daugman's normalization process. r and θ are the coordinates of non-concentric form. The number of points along a radial line is known as radial resolution (r), and the number of the radial line all through the iris region is known as angular resolution (θ).

16.3.3 Feature extraction and encoding

The most discriminating feature of an iris has to be extracted to get an accurate recognition system. The phase is the significant feature of the iris image. Most of the recognition systems make use of 1D log-Gabor band pass filter to decompose an iris image. Daugman's phase quantization method is used for feature encoding. 1D log-Gabor filter frequency response is shown in Eq. (16.2):

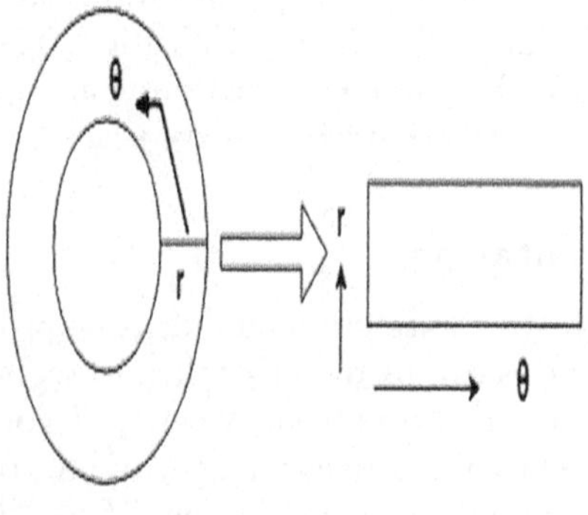

Figure 16.1 Outline of normalization.

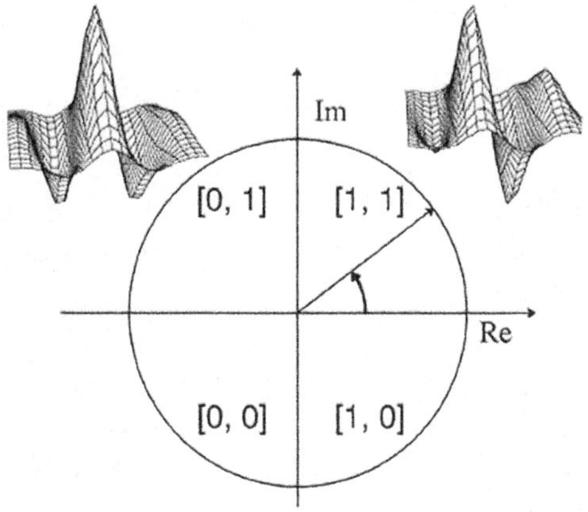

Figure 16.2 Daugman's phase quantization.

$$H(f) = \exp\left(\frac{-(\log(f / f_0))^2}{2(\log(\sigma / f_0))^2}\right), \tag{16.2}$$

where f_0 is center frequency of filter, σ decides the bandwidth of a filter used to maintain shape.

The filter output is phase quantized by Daugman's phase quantization method shown in Figure 16.2. The resultant template size = (2 × radial resolution × angular resolution).

16.4 PROPOSED METHOD

This work proposes implementation and analysis of a secure biometric authentication system based on iris. The basic block diagram depicting the architecture of the proposed system using the NTRU HE scheme is shown in Figure 16.3. Various processing tasks involved in biometric identification and verification system are carried out in encrypted domain to ensure data confidentiality and privacy together with efficiency and accuracy. The major three modules in this proposed biometric identification system are (i) feature extraction, (ii) encryption, (iii) secure domain matching. The feature extraction module is implemented using Daugman's feature extraction method as explained in Section 16.3.

The encryption scheme needs to be homomorphic to support classification in the encrypted domain. Fully homomorphic systems are highly computationally intensive causing a very high structural complexity. PHE schemes, on the other hand, provide lesser complexity solutions probably with lower accuracy. In this attempt, the NTRU encryption scheme is used

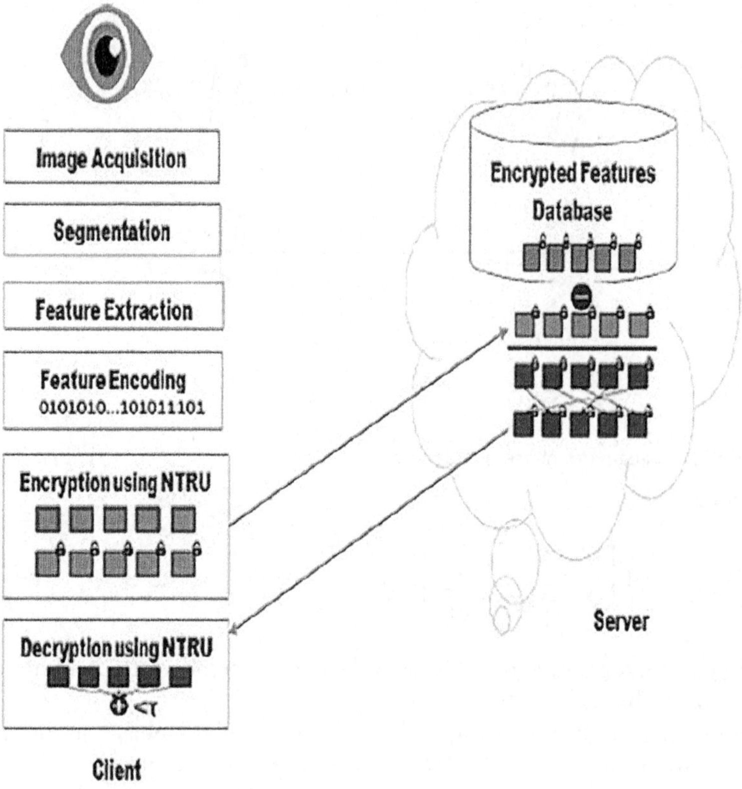

Figure 16.3 Architecture of the proposed secure domain biometric authentication system.

as PHE, since it supports feature matching in an encrypted domain with good accuracy levels and low complexity of operations.

16.4.1 NTRU encryption

NTRU [13] is a lattice-based cryptosystem which involves lattices in construction of algorithm. NTRU is a post-quantum cryptography based on a shortest-vector problem in a lattice. Since NTRU is described over convolutional polynomial rings which support two operations, it can be extended to FHE. Hoffstein, Pipher, and Silverman proposed an NTRU scheme. This algorithm is standardized in IEEE 1363.1 and it is highly safe as it is immune to the attacks by quantum computers.

Let (N, p, q) be the three integer public parameters in NTRU algorithm, where N is a degree of polynomials in a ring and a prime number. p and q are integers and $\gcd(p, q) = 1$. Always consider q as larger than p.

The parameters (y_1, y_2, y_3) are used for product key generation. The number of non-zero coefficients in private key component g is denoted by y_g, and y_m is a message hamming weight constraint. All the operations of NTRU scheme are performed in $z_q[x]/(x^N - 1)$ or in $z_p[x]/(x^N - 1)$. Let τ_N be a set of trinary polynomials, such that τ_N (y, e) be a set of ternary polynomials with exactly y ones and e number of minus ones in each polynomial, and $P(x_1, x_2, x_3)$ is a product form private key polynomial. The private key components can be considered as $(f, g) = (1 + pF, g)$ with $F \in P(y_1, y_2, y_3)$ and $g \in \tau_N (y_{g+} 1, y_g)$. The polynomial f has inverse modulo q and inverse modulo p denoted by fq and fp.

16.4.1.1 Advantages

- Smallest compiled code and consumes minimal CPU and battery resources.
- Particularly well suited for embedded and mobile devices where code size is a major limitation.
- Highest performing public key cryptography.
- Encryption and decryption of NTRU are faster than RSA at equal security level.

Algorithm 16.1 NTRUEncrypt key generation

Input: A full set of NTRUEncrypt parameters.
1. Repeat
2. $F \leftarrow P(x_1, x_2, x_3)$
3. $f = 1 + pF$
4. Until f invertible in R_N, q, and R_N, p
5. $g \leftarrow \tau_N (y_g + 1, y_g)$
6. Public key $h = fqg \in R_N$, p

Output: Public key h, private key (f, fp).

Algorithm 16.2 NTRUEncrypt encryption

Input: Public key h, message $M \in \{0,1\}$, and set the parameter $p = 3$.
1. Select the random polynomial r, which is used to obscure the message.
2. $r \leftarrow \tau_N (y_m + 1, y_m)$
3. Encryption $e = prh + M \pmod{q}$

Algorithm 16.3 NTRUEncrypt decryption

Input: Private key (f, fp), ciphertext $e \in R_N$, q, and set the parameter $p = 3$.
1. Multiply e with f and arrange the coefficients to have in a range $[-q/2, q/2]$

2. $b = {}^{*} e$ (modulo q)
3. $a = b$(modulo p)
4. Finally, $M = a * fp$(modulo p)

16.4.1.2 Parameter selection

The security level offered by NTRU highly depends on parameter selection. Through detailed analysis, the size of parameters required to offer different levels of security are identified [14] and updated in Table 16.1. This NTRU encryption protocol is implemented in Sage Math [15].

16.4.2 Proposed secure domain biometric authentication

In the proposed protocol during the enrolment phase, extracted iris feature templates are encrypted as blocks using the NTRU encryption scheme and stored in a database. The size of different parameters in NTRU scheme is chosen for sufficient level of accuracy based on details given in Table 16.2. Then the required parameters are generated using Algorithm 16.1. Then the features developed using Daugman's feature extraction method are encrypted by following the steps given in Algorithm 16.2. During the authentication phase, a new iris feature template is extracted from the client, the feature vector is divided into blocks, and each block is encrypted using the NTRU encryption scheme (Algorithm 16.2). These encrypted blocks are sent to a server for matching and authentication. The server matches these encrypted blocks against enrolled feature blocks in a database. A simple subtraction of blocks is used to generate matching metric, since the additive homomorphism of the NTRU is exploited in this protocol. Then the subtracted blocks are interchanged to hide the order of the blocks from the client. The subtracted blocks are further communicated to client as shown in Figure 16.3. The client decrypts the blocks using Algorithm 16.3. The number of non-zero coefficients in the decryption result denotes the Hamming weight of the matching metric, which is further compared with the threshold value for authentication.

Table 16.1 NTRU parameters for different security levels

Security level (bits)	p	Q	y_1	y_2	y_3	y_g	y_m
112	3	2048	8	8	6	133	101
128	3	2048	9	8	5	146	112
192	3	2048	10	10	8	197	158
256	3	2048	11	11	15	247	204

16.5 EXPERIMENTAL RESULTS

Simulation studies are conducted using MATLAB 2018b. CASIA database [16] of eye images is used for experimental evaluation of proposed protocol. Since CASIA [16] eye database does not have any specular reflection, iris segmentation is directly applied to the eye image. Figure 16.4 shows the result of segmentation step. The results of iris normalization and feature encoding are shown in Figures 16.5 and 16.6, respectively.

The extracted iris template is encrypted using NTRU encryption scheme. This NTRU encryption protocol is implemented in Sage Math [15]. Assume that a client has created an NTRU key pair where h is a public key and (f, fp) are private keys. Both server and client split the message sequence into blocks of length n, creating S blocks. Let $\mathbb{D}$ contain a polynomial set of $z_p[x]/(x^N - 1)$ with coefficients –1, 0, and 1. α_i and $\beta_i \in [1, S]$ denote the blocks of client and server messages, respectively.

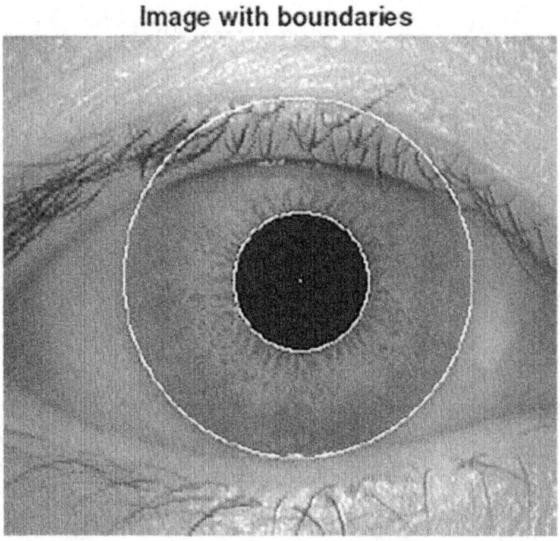

Figure 16.4 Iris segmentation.

Figure 16.5 Iris normalization.

Figure 16.6 Iris template.

Initially, the client sends the server the message

$$M_{A_i} = \{hr_i + \alpha_i\}, \forall\ i \in [1, s], \tag{16.3}$$

where r_i are random polynomials in $\mathbb{D}$. After receiving M_{A_i}'s, the server computes

$$M_{B_i} = \{M_{A_i} - hr_i' + \beta_i\}, \forall\ i \in [1, s], \tag{16.4}$$

where r_i' are random polynomials in $\mathbb{D}$. Then the server randomly changes the order of the blocks to hide from the client, and it is denoted by $M_{B_i}' \equiv \pi(M_{B_i})$, where π is a random permutation. Here we used the additive homomorphism of NTRU. After receiving M_{B_i}', client decrypts each M_{B_i}' and computes the total number of non-zero coefficients known as weight w_i of each recovered message. If $\sum_{i=1}^{s} w_i < \tau$, then the user authentication is accepted otherwise rejected. The threshold value τ is decided by plotting threshold versus error as shown in Figure 16.7. Various parameters chosen for simulation are as detailed below.

- Length of feature vector: 2,048 bits
- Length of each block: 64 bits

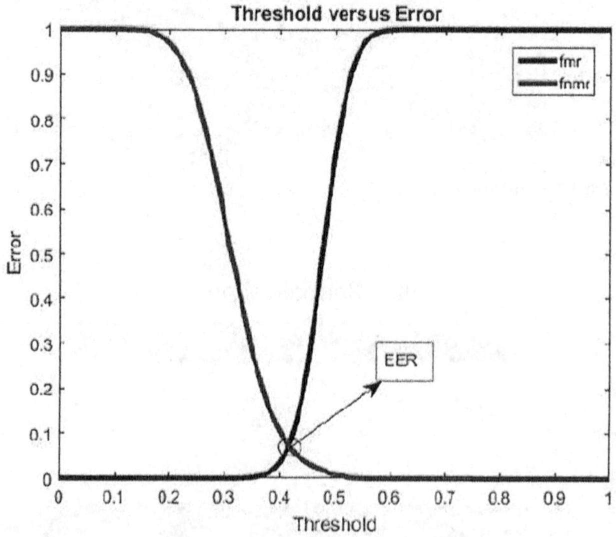

Figure 16.7 Threshold versus error.

Table 16.2 Time complexity of a protocol for single thread

Security level	*Key generation time*	*Client side (time for encryption and decryption of features)*	*Server side (time for encryption, comparison, and shuffling)*	*Total time*
128	0.1872	0.146	1.90445	2.237
192	0.3366	0.251	3.325	3.916
256	0.5696	0.3435	3.324	4.237

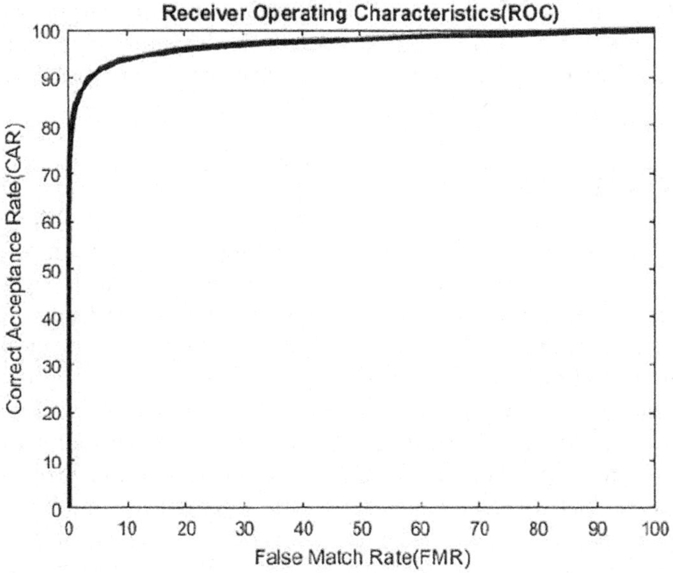

Figure 16.8 Receiver operating characteristic graph.

- Number of blocks: 32
- Resultant time is in seconds

The performance of the biometric verification system is evaluated in terms of two types of errors: (i) false match rate: used to measure the accuracy of the system in terms of rate of accepting an unauthorized person as authorized; and (ii) false non-match rate: used to measure the rate of rejecting the authorized person. The threshold in the proposed protocol is chosen to make these error rates minimum. The false match rate versus false non-match rate is shown in Figure 16.8, where threshold is fixed to 0.42 and equal error rate is 0.099. Receiver operating characteristic graph shows that the proposed system offers good accuracy in performance. In this experiment, total 324 images from CASIA database are used for classification giving 90% accuracy.

Table 16.3 Comparison of proposed protocol with FHE-based protocol

Method	*Key generation time*	*Encryption time*	*Decryption time*	*Total time*
Proposed method	0.1872 s	0.0428 s	0.2013 s	0.433 s
FHE-based method	26.649 s	3.8 min	0.49 s	3.27 min

The proposed protocol is compared with privacy-preserving biometric authentication system using FHE [17] for a comparable security level in Table 16.3. Since FHE allows both addition and multiplication operations on cipher text, this scheme can be implemented for any matching metric in encrypted domain. But the time complexity of this protocol is very high which makes it unsuitable for real-time implementation.

16.6 CONCLUSION

Biometric authentication helps to simplify user authentication and facilitates improved security in access control as biometric features cannot be reproduced by an unauthorized user. But it has privacy risks as the stored biometric features may be tampered by attackers with ulterior motives. To overcome this challenge, this work proposes a privacy-preserving biometric authentication protocol. HE preserves the privacy of biometric data because it supports authentication on encrypted data without sharing the private key to any other party during the authentication protocol. The proposed iris-based biometric authentication system uses additive homomorphic NTRU encryption scheme to enable secure, fast, and accurate privacy-preserving biometric authentication. The accuracy of the system is validated through simulation studies.

REFERENCES

1. M. A. M. Ali and N. M. Tahir. 2018. Cancellable biometrics technique for iris recognition. *IEEE Symposium on Computer Applications and Industrial Electronics (ISCAIE)*, Penang. 434–437.
2. N. K Ratha et al. 2001. Enhancing security and privacy in biometrics-based authentication systems. *IBM Systems Journal* 40, 614–634.
3. A. Goh et al. 2003. *Computation of Cryptographic Keys from Face Biometrics*. Springer Berlin Heidelberg, Berlin, Heidelberg.
4. A. Juels et al. 1999. A fuzzy commitment scheme. *Proceedings of the 6th ACM Conference on Computer and Communication Security*, ACM, New York. 28–36.
5. L. Yuan. 2014. Multimodal cryptosystem based on fuzzy commitment. *IEEE 17th International Conference on Computational Science and Engineering*, Chengdu. 1545–1549, IEEE.

6. Gerhard potzelsberger. 2013. KV Web Security – Applications of Homomorphic Encryption.
7. M. Yasuda et al. 2013. Packed Homomorphic Encryption based on Ideal Lattices and its Application to Biometrics. *Security Engineering and Intelligence Informatics*, Springer Berlin Heidelberg, Berlin, Heidelberg. 55–74.
8. W. A. A. Torres et al. 2015. Privacy preserving biometrics authentication systems using fully homomorphic encryption. *International Journal of Pervasive Computing and Communications*, 11. 151–168.
9. M. Naehrig et al. 2011. Can Homomorphic Encryption Be Practical? *Proceedings of the 3rd ACM Workshop on Cloud Computing Security Workshop, CCSW '11*, Association for Computing Machinery New York, NY, United States, 113–124.
10. C. Patsakis et al. 2016. Privacy-Preserving Biometric Authentication and Matching via Lattice Based Encryption. *Data Privacy Management, and Security Assurance*, Springer International Publishing, Verlag. 162–182.
11. J. Daugman, 2009. How Iris Recognition Works. In *The Essential Guide to Image Processing*, Academic Press, pp. 715–739.
12. P. Verma et al. 2012. Daughman's algorithm method for iris recognition—a biometric approach. *International Journal of Emerging Technology and Advanced Engineering*, 2(6), 177–185.
13. J. Hoffstein et al. 1998. NTRU: A ring-based public key cryptosystem. *Algorithmic Number Theory*, Springer Berlin Heidelberg, Berlin, Heidelberg, 267–288.
14. J. Hoffstein et al. 2017. *Choosing Parameters for NTRUEncrypt.* In *Cryptographers' Track at the RSA Conference* (pp. 3–18). Springer, Cham.
15. M. A. Gray. 2008. Sage: A new mathematics software system. *Computing in Science Engineering*. 10(6), 72–75.
16. A Dataset of 756 grey scale eye images; Chinese Academy of Sciences-Institute of Automation, http://www.sinobiometrics.com, Version 4.0.
17. W. A. A. Torres et al. 2015. Privacy preserving biometrics authentication systems using fully homomorphic encryption. *International Journal of Pervasive Computing and Communications*, 11. 151–168.

Index